SpringerBriefs in Mathematics

SpringerBriefs in Mathematics showcases expositions in all areas of mathematics and applied mathematics. Manuscripts presenting new results or a single new result in a classical field, new field, or an emerging topic, applications, or bridges between new results and already published works, are encouraged. The series is intended for mathematicians and applied mathematicians.

For further volumes:
http://www.springer.com/series/10030

Silvestru Sever Dragomir

Inequalities for the Numerical Radius of Linear Operators in Hilbert Spaces

 Springer

Silvestru Sever Dragomir
College of Engineering and Science
Victoria University
Melbourne, Australia

School of Computational
  and Applied Mathematics
University of the Withwatersrand
Braamfontein
Johannesburg, South Africa

ISSN 2191-8198   ISSN 2191-8201 (electronic)
ISBN 978-3-319-01447-0   ISBN 978-3-319-01448-7 (eBook)
DOI 10.1007/978-3-319-01448-7
Springer Cham Heidelberg New York Dordrecht London

Library of Congress Control Number: 2013946653

Mathematics Subject Classification (2010): 47A63, 47A12, 47A30, 47A05

Printed on acid-free paper

Springer is part of Springer Science+Business Media (www.springer.com)

To my granddaughter Audrey Elise

Preface

As pointed out by Gustafson and Rao in their seminal book [*Numerical Range. The Field of Values of Linear Operators and Matrices.* Universitext. Springer-Verlag, New York, 1997. xiv+189 pp.], the concepts of *numerical range* and *numerical radius* play an important role in various fields of contemporary mathematics, including operator theory, operator trigonometry, numerical analysis, and fluid dynamics.

Since 1997 the research devoted to these mathematical objects has grown greatly. A simple search in the database *MathSciNet* of the *American Mathematical Society* with the keyword "numerical range" in the title reveals more than 300 papers published after 1997 while the same search with the keyword "numerical radius" adds other 100, showing an immense interest on the subject by numerous researchers working in different fields of modern mathematics. If no restrictions for the year are imposed, the number of papers with those keywords in the title exceeds 1,000. However, the size of the areas of applications for numerical ranges and radii is very difficult to estimate. If we perform a search looking for the publications where in a way or another the concept of "numerical range" is used, we can get more than 1,550 items.

The present monograph is focused on numerical radius inequalities for bounded linear operators on complex Hilbert spaces for the case of one and two operators.

The book is intended for use both by researchers in various fields of linear operator theory in Hilbert spaces and mathematical inequalities, domains which have grown exponentially in the last decade, and by postgraduate students and scientists applying inequalities in their specific areas.

In the introductory chapter we present some fundamental facts about the numerical range and the numerical radius of bounded linear operators in Hilbert spaces. Some classical inequalities due to Berger, Holbrook, Fong and Holbrook and Bouldin are given. More recent and interesting results obtained by Kittaneh, El-Haddad and Kittanek and Yamazaki are provided as well.

In Chap. 2, we present recent results obtained by the author concerning numerical radius and norm inequalities for one operator on a complex Hilbert space. The techniques employed to prove the results are elementary. We also use some special vector inequalities in inner product spaces due to Buzano, Goldstein, Ryff and Clarke as well as some reverse Schwarz inequalities and Grüss type inequalities obtained by the author. Numerous references for the Kantorovich inequality that is extended to larger classes of operators than positive operators are provided as well.

In Chap. 3, we present recent results obtained by the author concerning the norms and the numerical radii of two bounded linear operators. The techniques in this case are also elementary and can be understood by undergraduate students taking a subject in operator theory. Some vector inequalities in inner product spaces as well as inequalities for means of nonnegative real numbers are also employed.

For the sake of completeness, all the results presented are completely proved and the original references where they have been firstly obtained are mentioned. The chapters are followed by the list of references used therein and therefore are relatively independent and can be read separately.

Melbourne, Australia   Silvestru Sever Dragomir

Contents

Chapter 1
Introduction

In this introductory chapter we present some fundamental facts about the numerical range and the numerical radius of bounded linear operators in Hilbert spaces that are used throughout the book. Some famous inequalities due to Berger, Holbrook, Fong and Holbrook and Bouldin are given. More recent results obtained by Kittaneh, El-Haddad and Kittanek and Yamazaki are provided as well.

1.1 Basic Definitions and Facts

Let $(H; \langle \cdot, \cdot \rangle)$ be a complex Hilbert space. The *numerical range* of an operator T is the subset of the complex numbers $\mathbb{C}$ given by [7, p. 1]:

$$W(T) = \{\langle Tx, x \rangle, \ x \in H, \ \|x\| = 1\}.$$

The following properties of $W(T)$ are immediate:

(i) $W(\alpha I + \beta T) = \alpha + \beta W(T)$ for $\alpha, \beta \in \mathbb{C}$;
(ii) $W(T^*) = \{\bar{\lambda}, \lambda \in W(T)\}$, where T^* is the *adjoint operator* of T;
(iii) $W(U^*TU) = W(T)$ for any *unitary* operator U.

The following classical fact about the geometry of the numerical range [7, p. 4] may be stated:

Theorem 1 (Toeplitz-Hausdorff). *The numerical range of an operator is convex.*

An important use of $W(T)$ is to bound the *spectrum* $\sigma(T)$ of the operator T [7, p. 6]:

Theorem 2 (Spectral Inclusion). *The spectrum of an operator is contained in the closure of its numerical range.*

S.S. Dragomir, *Inequalities for the Numerical Radius of Linear Operators in Hilbert Spaces*, SpringerBriefs in Mathematics, DOI 10.1007/978-3-319-01448-7_1, © Silvestru Sever Dragomir 2013

The self-adjoint operators have their spectra bounded sharply by the numerical range [7, p. 7]:

Theorem 3. *The following statements hold true:*

(i) T *is self-adjoint iff* $W(T)$ *is real;*
(ii) *If* T *is self-adjoint and* $W(T) = [m, M]$ *(the closed interval of real numbers* m, M *), then* $\|T\| = \max\{|m|, |M|\}$;
(iii) *If* $W(T) = [m, M]$, *then* $m, M \in \sigma(T)$.

The *numerical radius* $w(T)$ of an operator T on H is given by [7, p. 8]:

$$w(T) = \sup\{|\lambda|, \lambda \in W(T)\} = \sup\{|\langle Tx, x\rangle|, \|x\| = 1\}. \qquad (1.1)$$

Obviously, by (1.1), for any $x \in H$, one has

$$|\langle Tx, x\rangle| \le w(T)\|x\|^2. \qquad (1.2)$$

It is well known that $w(\cdot)$ is a norm on the Banach algebra $B(H)$ of all bounded linear operators $T : H \to H$, i.e.,

(i) $w(T) \ge 0$ for any $T \in B(H)$ and $w(T) = 0$ if and only if $T = 0$;
(ii) $w(\lambda T) = |\lambda| w(T)$ for any $\lambda \in \mathbb{C}$ and $T \in B(H)$;
(iii) $w(T + V) \le w(T) + w(V)$ for any $T, V \in B(H)$.

This norm is equivalent with the operator norm. In fact, the following more precise result holds [7, p. 9]:

Theorem 4 (Equivalent Norm). *For any* $T \in B(H)$ *one has*

$$w(T) \le \|T\| \le 2w(T). \qquad (1.3)$$

Let us now look at two extreme cases of the inequality (1.3). In the following $r(t) := \sup\{|\lambda|, \lambda \in \sigma(T)\}$ will denote the *spectral radius* of T and $\sigma_p(T) = \{\lambda \in \sigma(T), Tf = \lambda f \text{ for some } f \in H\}$ the *point spectrum* of T.
The following results hold [7, p. 10]:

Theorem 5. *We have*

(i) *If* $w(T) = \|T\|$, *then* $r(T) = \|T\|$;
(ii) *If* $\lambda \in W(T)$ *and* $|\lambda| = \|T\|$, *then* $\lambda \in \sigma_p(T)$.

To address the other extreme case $w(T) = \frac{1}{2}\|T\|$, we can state the following sufficient condition in terms of (see [7, p. 11])

$$R(T) := \{Tf, f \in H\} \quad \text{and} \quad R(T^*) := \{T^*f, f \in H\}.$$

Theorem 6. *If* $R(T) \perp R(T^*)$, *then* $w(T) = \frac{1}{2}\|T\|$.

It is well known that the two-dimensional shift

$$S_2 = \begin{bmatrix} 0 & 0 \\ 1 & 0 \end{bmatrix},$$

has the property that $w(T) = \frac{1}{2}\|T\|$.

The following theorem shows that some operators T with $w(T) = \frac{1}{2}\|T\|$ have S_2 as a component [7, p. 11]:

Theorem 7. *If $w(T) = \frac{1}{2}\|T\|$ and T attains its norm, then T has a two-dimensional reducing subspace on which it is the shift S_2.*

For other results on numerical radius, see [8], Chap. 11.

1.2 Results for One Operator

The following power inequality for one operator is a classical result in the field (for a simple proof see [14]):

Theorem 8 (Berger [2], 1965). *For any operator $T \in B(H)$ and natural number n we have*

$$w(T^n) \le w^n(T).$$

Further, we list some recent inequalities for one operator.

Theorem 9 (Kittaneh [10], 2003). *For any operator $T \in B(H)$ we have the following refinement of the first inequality in (1.3)*

$$w(T) \le \frac{1}{2}\left(\|T\| + \|T^2\|^{1/2}\right). \tag{1.4}$$

Utilizing the Cartesian decomposition for operators, F. Kittaneh improves the inequality (1.3) as follows:

Theorem 10 (Kittaneh [11], 2005). *For any operator $T \in B(H)$ we have*

$$\frac{1}{4}\|T^*T + TT^*\| \le w^2(T) \le \frac{1}{2}\|T^*T + TT^*\|. \tag{1.5}$$

When more information concerning the angle between the ranges of T and T^* is available, the following interesting estimate holds:

Theorem 11 (Bouldin [3], 1971). *If we denote by α the angle between the ranges of T and T^*, then*

$$w(T) \le \frac{1}{2}\|T\|\left[\cos\alpha + \left(\cos^2\alpha + 1\right)^{1/2}\right]. \tag{1.6}$$

For powers of the absolute value of operators, one can state the following results:

Theorem 12 (El-Haddad and Kittaneh [5], 2007). *If for an operator $T \in B(H)$ we denote $|T| := (T^*T)^{1/2}$, then*

$$w^r(T) \le \frac{1}{2}\left\||T|^{2\alpha r} + |T^*|^{2(1-\alpha)r}\right\| \tag{1.7}$$

and

$$w^{2r}(T) \le \left\| \alpha \, |T|^{2r} + (1-\alpha) \, |T^*|^{2r} \right\| \tag{1.8}$$

where $\alpha \in (0,1)$ and $r \ge 1$.

If we take $\alpha = \frac{1}{2}$ and $r = 1$ we get from (1.7)

$$w(T) \le \frac{1}{2} \left\| |T| + |T^*| \right\| \tag{1.9}$$

and from (1.8)

$$w^2(T) \le \frac{1}{2} \left\| |T|^2 + |T^*|^2 \right\|. \tag{1.10}$$

For the Cartesian decomposition of T we have:

Theorem 13 (El-Haddad and Kittaneh [5], 2007). *If $T = B + iC$ is the Cartesian decomposition of T, then*

$$w^r(T) \le \left\| |B|^r + |C|^r \right\| \tag{1.11}$$

for $r \in (0,2]$.

If $r \ge 2$, *then*

$$w^r(T) \le 2^{\frac{r}{2}-1} \left\| |B|^r + |C|^r \right\| \tag{1.12}$$

and

$$2^{-\frac{r}{2}-1} \left\| |B+C|^r + |B-C|^r \right\| \tag{1.13}$$

$$\le w^r(T) \le \frac{1}{2} \left\| |B+C|^r + |B-C|^r \right\|.$$

We observe that for $r = 1$ we get from (1.11)

$$w(T) \le \left\| |B| + |C| \right\| \tag{1.14}$$

while for $r = 2$ we get from (1.12) or from (1.11)

$$w^2(T) \le \left\| |B|^2 + |C|^2 \right\| \tag{1.15}$$

and from (1.13)

$$\frac{1}{4} \left\| |B+C|^2 + |B-C|^2 \right\| \le w^2(T) \le \frac{1}{2} \left\| |B+C|^2 + |B-C|^2 \right\|. \tag{1.16}$$

Let $T = U\,|T|$ be the *polar decomposition* of the bounded linear operator T. The *Aluthge transform* $\tilde{T}$ of T is defined by $\tilde{T} := |T|^{1/2} U \, |T|^{1/2}$, see [1].

The following properties of $\tilde{T}$ are as follows:

(i) $\|\tilde{T}\| \le \|T\|$;
(ii) $w\left(\tilde{T}\right) \le w\left(T\right)$;
(iii) $r\left(\tilde{T}\right) = w\left(T\right)$;
(iv) $w\left(\tilde{T}\right) \le \left\|T^2\right\|^{1/2} \left(\le \|T\|\right)$ [15].

Utilizing this transform one can obtain the following refinement of Kittaneh's inequality (1.4).

Theorem 14 (Yamazaki [15], 2007). *For any operator* $T \in B\left(H\right)$ *we have*

$$w\left(T\right) \le \frac{1}{2}\left(\|T\| + w\left(\tilde{T}\right)\right) \le \frac{1}{2}\left(\|T\| + \left\|T^2\right\|^{1/2}\right). \qquad (1.17)$$

We remark that if $\tilde{T} = 0$, then obviously $w\left(T\right) = \frac{1}{2}\|T\|$.

1.3   Results for Two Operators

The following general result for the product of two operators holds [7, p. 37]:

Theorem 15 (Holbrook [9], 1969). *If* A, B *are two bounded linear operators on the Hilbert space* $\left(H, \langle\cdot,\cdot\rangle\right)$, *then* $w\left(AB\right) \le 4w\left(A\right)w\left(B\right)$. *In the case that* $AB = BA$, *then* $w\left(AB\right) \le 2w\left(A\right)w\left(B\right)$. *The constant 2 is best possible here.*

The following results are also well known [7, p. 38].

Theorem 16 (Holbrook [9], 1969). *If* A *is a unitary operator that commutes with another operator* B, *then*

$$w\left(AB\right) \le w\left(B\right). \qquad (1.18)$$

If A *is an isometry and* $AB = BA$, *then (1.18) also holds true.*

We say that A and B *double commute* if $AB = BA$ and $AB^* = B^*A$. The following result holds [7, p. 38].

Theorem 17 (Holbrook [9], 1969). *If the operators* A *and* B *double commute, then*

$$w\left(AB\right) \le w\left(B\right)\|A\|. \qquad (1.19)$$

As a consequence of the above, we have [7, p. 39]:

Corollary 18. *Let* A *be a normal operator commuting with* B. *Then*

$$w\left(AB\right) \le w\left(A\right)w\left(B\right). \qquad (1.20)$$

A related problem with the inequality (1.19) is to find the best constant c for which the inequality

$$w(AB) \leq cw(A)\|B\|$$

holds for any two commuting operators $A, B \in B(H)$. It is known that $1.064 < c < 1.169$, see [4, 12, 13].

In relation to this problem, it has been shown that:

Theorem 19 (Fong and Holbrook [6], 1983). *For any $A, B \in B(H)$ we have*

$$w(AB + BA) \leq 2\sqrt{2}w(A)\|B\|. \tag{1.21}$$

The following result for several operators holds:

Theorem 20 (Kittaneh [11], 2005). *For any $A, B, C, D, S, T \in B(H)$ we have*

$$w(ATB + CSD)$$

$$\leq \frac{1}{2}\left\| A\,|T^*|^{2(1-\alpha)}\,A^* + B^*\,|T|^{2\alpha}\,B + C\,|S^*|^{2(1-\alpha)}\,C^* + D^*\,|T|^{2\alpha}\,D \right\|, \tag{1.22}$$

where $\alpha \in [0, 1]$.

Following [11] we list here some particular inequalities of interest.

If we take $T = I$ and $S = 0$ in (1.22) we get

$$w(AB) \leq \frac{1}{2}\|AA^* + B^*B\|. \tag{1.23}$$

In addition to this we have the related inequality

$$w(AB) \leq \frac{1}{2}\|A^*A + BB^*\|. \tag{1.24}$$

If we choose $T = S = I$, $C = B$ and $D = \pm A$ in (1.22) we get

$$w(AB \pm BA) \leq \frac{1}{2}\|A^*A + AA^* + BB^* + B^*B\| \tag{1.25}$$

which provides an upper bound for the numerical radius of the commutator $AB - BA$.

If we take $\alpha = \frac{1}{2}$ in (1.22) we also can derive the inequality

$$w\left(AB \pm B^*A\right) \leq \frac{1}{2}\left\| |A| + |A^*| + B^*\left(|A| + |A^*|\right)B \right\|. \tag{1.26}$$

References

1. Aluthge, A.: Some generalized theorems on p-hyponormal operators. Integral Equ. Operator Theory **24**, 497–501 (1996)
2. Berger, C.: A strange dilation theorem. Notices Am. Math. Soc. **12**, 590 (1965) [Abstract 625–152]
3. Bouldin, R.: The numerical range of a product. II. J. Math. Anal. Appl. **33**, 212–219 (1971)
4. Davidson, K.R., Holbrook, J.A.R.: Numerical radii of zero-one matricies. Michigan Math. J. **35**, 261–267 (1988)
5. El-Haddad, M., Kittaneh, F.: Numerical radius inequalities for Hilbert space operators. II. Studia Math. **182**(2), 133–140 (2007)
6. Fong, C.K., Holbrook, J.A.R.: Unitarily invariant operators norms. Canad. J. Math. **35**, 274–299 (1983)
7. Gustafson, K.E., Rao, D.K.M.: Numerical Range. Springer, New York, Inc. (1997)
8. Halmos, P.R.: A Hilbert Space Problem Book, 2nd edn. Springer, New York (1982)
9. Holbrook, J.A.R.: Multiplicative properties of the numerical radius in operator theory. J. Reine Angew. Math. **237**, 166–174 (1969)
10. Kittaneh, F.: A numerical radius inequality and an estimate for the numerical radius of the Frobenius companion matrix. Studia Math. **158**(1), 11–17 (2003)
11. Kittaneh, F.: Numerical radius inequalities for Hilbert space operators. Studia Math. **168**(1), 73–80 (2005)
12. Müller, V.: The numerical radius of a commuting product. Michigan Math. J. **39**, 255–260 (1988)
13. Okubo, K., Ando, T.: Operator radii of commuting products. Proc. Am. Math. Soc. **56**, 203–210 (1976)
14. Pearcy, C.: An elementary proof of the power inequality for the numerical radius. Michigan Math. J. **13**, 289–291 (1966)
15. Yamazaki, T. On upper and lower bounds of the numerical radius and an equality condition. Studia Math. **178**(1), 83–89 (2007)

Chapter 2
Inequalities for One Operator

In this chapter we present with complete proofs some recent results obtained by the author concerning numerical radius and norm inequalities for a bounded linear operator on a complex Hilbert space. The techniques employed to prove the results are elementary. We also use some special vector inequalities in inner product spaces due to Buzano, Goldstein, Ryff and Clarke, Dragomir and Sándor as well as some reverse Schwarz inequalities and Grüss type inequalities obtained by the author. Many references for the Kantorovich inequality that is extended here to larger classes of operators than positive operators are provided as well.

2.1 Reverse Inequalities for the Numerical Radius

2.1.1 Reverse Inequalities

The following results may be stated:

Theorem 21 (Dragomir [13], 2005). *Let* $T : H \to H$ *be a bounded linear operator on the complex Hilbert space* H. *If* $\lambda \in \mathbb{C} \backslash \{0\}$ *and* $r > 0$ *are such that*

$$\|T - \lambda I\| \leq r, \tag{2.1}$$

where $I : H \to H$ *is the identity operator on* H, *then*

$$(0 \leq) \|T\| - w(T) \leq \frac{1}{2} \cdot \frac{r^2}{|\lambda|}. \tag{2.2}$$

Proof. For $x \in H$ with $\|x\| = 1$, we have from (2.1) that

$$\|Tx - \lambda x\| \leq \|T - \lambda I\| \leq r,$$

S.S. Dragomir, *Inequalities for the Numerical Radius of Linear Operators in Hilbert Spaces*, SpringerBriefs in Mathematics, DOI 10.1007/978-3-319-01448-7_2, © Silvestru Sever Dragomir 2013

giving

$$\|Tx\|^2 + |\lambda|^2 \le 2\,\mathrm{Re}\left[\overline{\lambda}\,\langle Tx, x\rangle\right] + r^2 \tag{2.3}$$

$$\le 2\,|\lambda|\,|\langle Tx, x\rangle| + r^2.$$

Taking the supremum over $x \in H$, $\|x\| = 1$ in (2.3) we get the following inequality that is of interest in itself:

$$\|T\|^2 + |\lambda|^2 \le 2w\,(T)\,|\lambda| + r^2. \tag{2.4}$$

Since, obviously,

$$2\,\|T\|\,|\lambda| \le \|T\|^2 + |\lambda|^2, \tag{2.5}$$

hence by (2.4) and (2.5) we deduce the desired inequality (2.2). ∎

Remark 22. If the operator $T : H \to H$ is such that $R\,(T) \perp R\,(T^*)$, $\|T\| = 1$ and $\|T - I\| \le 1$, then the equality holds in (2.2). Indeed, by Theorem 6, we have in this case $w\,(T) = \frac{1}{2}\,\|T\| = \frac{1}{2}$ and since we can choose $\lambda = 1$, $r = 1$ in Theorem 21, then we get in both sides of (2.2) the same quantity $\frac{1}{2}$.

The following corollary may be stated:

Corollary 23. *Let $T : H \to H$ be a bounded linear operator and $\varphi, \psi \in \mathbb{C}$ with $\psi \notin \{-\varphi, \varphi\}$. If*

$$\mathrm{Re}\,\langle \psi x - Tx, Tx - \varphi x\rangle \ge 0 \quad \text{for any} \quad x \in H, \ \|x\| = 1 \tag{2.6}$$

then

$$(0 \le)\,\|T\| - w\,(T) \le \frac{1}{4} \cdot \frac{|\psi - \varphi|^2}{|\psi + \varphi|}. \tag{2.7}$$

Proof. Utilizing the fact that in any Hilbert space the following two statements are equivalent:

(i) $\mathrm{Re}\,\langle u - x, x - z\rangle \ge 0$, $x, z, u \in H$;
(ii) $\left\|x - \frac{z+u}{2}\right\| \le \frac{1}{2}\,\|u - z\|$,

we deduce that (2.6) is equivalent to

$$\left\|Tx - \frac{\psi + \varphi}{2} \cdot I x\right\| \le \frac{1}{2}\,|\psi - \varphi| \tag{2.8}$$

for any $x \in H$, $\|x\| = 1$, which in its turn is equivalent to the operator norm inequality:

$$\left\|T - \frac{\psi + \varphi}{2} \cdot I\right\| \le \frac{1}{2}\,|\psi - \varphi|. \tag{2.9}$$

Now, applying Theorem 21 for $T = T$, $\lambda = \frac{\varphi + \psi}{2}$ and $r = \frac{1}{2}\,|\Gamma - \gamma|$, we deduce the desired result (2.7). ∎

Remark 24. Following [33, p. 25], we say that an operator $B : H \rightarrow H$ is accretive, if $\operatorname{Re} \langle Bx, x \rangle \geq 0$ for any $x \in H$. One may observe that the assumption (2.6) above is then equivalent with the fact that the operator $(T^* - \bar{\varphi}I)(\psi I - T)$ is accretive.

Perhaps a more convenient sufficient condition in terms of positive operators is the following one:

Corollary 25. *Let* $\varphi, \psi \in \mathbb{C}$ *with* $\psi \notin \{-\varphi, \varphi\}$ *and* $T : H \rightarrow H$ *a bounded linear operator in* H. *If* $(T^* - \bar{\varphi}I)(\psi I - T)$ *is self-adjoint and*

$$\left(T^* - \bar{\varphi}I\right)\left(\psi I - T\right) \geq 0 \tag{2.10}$$

in the operator partial order, then

$$(0 \leq) \|T\| - w(T) \leq \frac{1}{4} \cdot \frac{|\psi - \varphi|^2}{|\psi + \varphi|}. \tag{2.11}$$

Corollary 26. *Assume that* T, λ, r *are as in Theorem 21. If, in addition,*

$$||\lambda| - w(T)| \geq \rho, \tag{2.12}$$

for some $\rho \geq 0$, *then*

$$(0 \leq) \|T\|^2 - w^2(T) \leq r^2 - \rho^2. \tag{2.13}$$

Proof. From (2.4) of Theorem 21, we have

$$\|T\|^2 - w^2(T) \leq r^2 - w^2(T) + 2w(T)|\lambda| - |\lambda|^2$$
$$= r^2 - (|\lambda| - w(T))^2. \tag{2.14}$$

The desired inequality follows from (2.12). ∎

Remark 27. In particular, if $\|T - \lambda I\| \leq r$ and $|\lambda| = w(T)$, $\lambda \in \mathbb{C}$, then

$$(0 \leq) \|T\|^2 - w^2(T) \leq r^2. \tag{2.15}$$

The following result may be stated as well.

Theorem 28 (Dragomir [13], 2005). *Let* $T : H \rightarrow H$ *be a nonzero bounded linear operator on* H *and* $\lambda \in \mathbb{C} \setminus \{0\}$, $r > 0$ *with* $|\lambda| > r$. *If*

$$\|T - \lambda I\| \leq r, \tag{2.16}$$

then

$$\sqrt{1 - \frac{r^2}{|\lambda|^2}} \leq \frac{w(T)}{\|T\|} \quad (\leq 1). \tag{2.17}$$

Proof. From (2.4) of Theorem 21, we have

$$\|T\|^2 + |\lambda|^2 - r^2 \le 2 |\lambda| w(T),$$

which implies, on dividing with $\sqrt{|\lambda|^2 - r^2} > 0$ that

$$\frac{\|T\|^2}{\sqrt{|\lambda|^2 - r^2}} + \sqrt{|\lambda|^2 - r^2} \le \frac{2 |\lambda| w(T)}{\sqrt{|\lambda|^2 - r^2}}. \qquad (2.18)$$

By the elementary inequality

$$2 \|T\| \le \frac{\|T\|^2}{\sqrt{|\lambda|^2 - r^2}} + \sqrt{|\lambda|^2 - r^2} \qquad (2.19)$$

and by (2.18) we deduce

$$\|T\| \le \frac{w(T) |\lambda|}{\sqrt{|\lambda|^2 - r^2}},$$

which is equivalent to (2.17).   ∎

Remark 29. Squaring (2.17), we get the inequality

$$(0 \le) \|T\|^2 - w^2(T) \le \frac{r^2}{|\lambda|^2} \|T\|^2. \qquad (2.20)$$

Remark 30. For any bounded linear operator $T : H \to H$ we have the relation $w(T) \ge \frac{1}{2} \|T\|$. Inequality (2.17) would produce a refinement of this classic fact only in the case when

$$\frac{1}{2} \le \left(1 - \frac{r^2}{|\lambda|^2} \right)^{\frac{1}{2}},$$

which is equivalent to $r / |\lambda| \le \sqrt{3}/2$.

The following corollary holds.

Corollary 31. *Let $\varphi, \psi \in \mathbb{C}$ with $\mathrm{Re}(\psi \bar{\varphi}) > 0$. If $T : H \to H$ is a bounded linear operator such that either (2.6) or (2.10) holds true, then:*

$$\frac{2\sqrt{\mathrm{Re}(\psi \bar{\varphi})}}{|\psi + \varphi|} \le \frac{w(T)}{\|T\|} \ (\le 1) \qquad (2.21)$$

and

$$(0 \leq) \|T\|^2 - w^2(T) \leq \left| \frac{\psi - \varphi}{\psi + \varphi} \right|^2 \|T\|^2 . \qquad (2.22)$$

Proof. If we consider $\lambda = \frac{\psi + \varphi}{2}$ and $r = \frac{1}{2} |\psi - \varphi|$, then

$$|\lambda|^2 - r^2 = \left| \frac{\psi + \varphi}{2} \right|^2 - \left| \frac{\psi - \varphi}{2} \right|^2 = \mathrm{Re}\,(\psi \bar{\varphi}) > 0.$$

Now, on applying Theorem 28, we deduce the desired result. ∎

Remark 32. If $|\psi - \varphi| \leq \frac{\sqrt{3}}{2} |\psi + \varphi|$, $\mathrm{Re}\,(\psi \bar{\varphi}) > 0$, then (2.21) is a refinement of the inequality $w(T) \geq \frac{1}{2} \|T\|$.

The following result may be of interest as well.

Theorem 33 (Dragomir [13], 2005). *Let* $T : H \to H$ *be a nonzero bounded linear operator on H and $\lambda \in \mathbb{C} \setminus \{0\}$, $r > 0$ with $|\lambda| > r$. If*

$$\|T - \lambda I\| \leq r, \qquad (2.23)$$

then

$$(0 \leq) \|T\|^2 - w^2(T) \leq \frac{2r^2}{|\lambda| + \sqrt{|\lambda|^2 - r^2}} w(T) . \qquad (2.24)$$

Proof. From the proof of Theorem 21, we have

$$\|Tx\|^2 + |\lambda|^2 \leq 2 \mathrm{Re} \left[\bar{\lambda} \langle Tx, x \rangle \right] + r^2 \qquad (2.25)$$

for any $x \in H$, $\|x\| = 1$.

If we divide (2.25) by $|\lambda| |\langle Tx, x \rangle|$, (which, by (2.25), is positive), then we obtain

$$\frac{\|Tx\|^2}{|\lambda| |\langle Tx, x \rangle|} \leq \frac{2 \mathrm{Re} \left[\bar{\lambda} \langle Tx, x \rangle \right]}{|\lambda| |\langle Tx, x \rangle|} + \frac{r^2}{|\lambda| |\langle Tx, x \rangle|} - \frac{|\lambda|}{|\langle Tx, x \rangle|} \qquad (2.26)$$

for any $x \in H$, $\|x\| = 1$.

If we subtract in (2.26) the same quantity $\frac{|\langle Tx, x \rangle|}{|\lambda|}$ from both sides, then we get

$$\frac{\|Tx\|^2}{|\lambda| |\langle Tx, x \rangle|} - \frac{|\langle Tx, x \rangle|}{|\lambda|}$$

$$\leq \frac{2 \mathrm{Re} \left[\bar{\lambda} \langle Tx, x \rangle \right]}{|\lambda| |\langle Tx, x \rangle|} + \frac{r^2}{|\lambda| |\langle Tx, x \rangle|} - \frac{|\langle Tx, x \rangle|}{|\lambda|} - \frac{|\lambda|}{|\langle Tx, x \rangle|}$$

$$= \frac{2\operatorname{Re}\left[\bar{\lambda}\langle Tx,x\rangle\right]}{|\lambda||\langle Tx,x\rangle|} - \frac{|\lambda|^2 - r^2}{|\lambda||\langle Tx,x\rangle|} - \frac{|\langle Tx,x\rangle|}{|\lambda|}$$

$$= \frac{2\operatorname{Re}\left[\bar{\lambda}\langle Tx,x\rangle\right]}{|\lambda||\langle Tx,x\rangle|} - \left(\frac{\sqrt{|\lambda|^2 - r^2}}{\sqrt{|\lambda||\langle Tx,x\rangle|}} - \frac{\sqrt{|\langle Tx,x\rangle|}}{\sqrt{|\lambda|}}\right)^2$$

$$- 2\frac{\sqrt{|\lambda|^2 - r^2}}{|\lambda|}. \tag{2.27}$$

Since

$$\operatorname{Re}\left[\bar{\lambda}\langle Tx,x\rangle\right] \le |\lambda||\langle Tx,x\rangle|$$

and

$$\left(\frac{\sqrt{|\lambda|^2 - r^2}}{\sqrt{|\lambda||\langle Tx,x\rangle|}} - \frac{\sqrt{|\langle Tx,x\rangle|}}{\sqrt{|\lambda|}}\right)^2 \ge 0,$$

by (2.27) we get

$$\frac{\|Tx\|^2}{|\lambda||\langle Tx,x\rangle|} - \frac{|\langle Tx,x\rangle|}{|\lambda|} \le \frac{2\left(|\lambda| - \sqrt{|\lambda|^2 - r^2}\right)}{|\lambda|}$$

which gives the inequality

$$\|Tx\|^2 \le |\langle Tx,x\rangle|^2 + 2|\langle Tx,x\rangle|\left(|\lambda| - \sqrt{|\lambda|^2 - r^2}\right) \tag{2.28}$$

for any $x \in H$, $\|x\| = 1$.

Taking the supremum over $x \in H$, $\|x\| = 1$, we get

$$\|T\|^2 \le \sup\left\{|\langle Tx,x\rangle|^2 + 2|\langle Tx,x\rangle|\left(|\lambda| - \sqrt{|\lambda|^2 - r^2}\right)\right\}$$

$$\le \sup\left\{|\langle Tx,x\rangle|^2\right\} + 2\left(|\lambda| - \sqrt{|\lambda|^2 - r^2}\right)\sup\{|\langle Tx,x\rangle|\}$$

$$= w^2(T) + 2\left(|\lambda| - \sqrt{|\lambda|^2 - r^2}\right)w(T),$$

which is clearly equivalent to (2.24).   ∎

Corollary 34. *Let $\varphi, \psi \in \mathbb{C}$ with $\operatorname{Re}(\psi\bar{\varphi}) > 0$. If $T : H \to H$ is a bounded linear operator such that either (2.6) or (2.10) hold true, then:*

$$(0 \le) \|T\|^2 - w^2(T) \le \left[|\psi + \varphi| - 2\sqrt{\operatorname{Re}(\psi\bar{\varphi})}\right] w(T). \qquad (2.29)$$

Remark 35. If $M \ge m > 0$ are such that either $(T^* - mI)(MI - T)$ is accretive, or, sufficiently, $(T^* - mI)(MI - T)$ is self-adjoint and

$$(T^* - mI)(MI - T) \ge 0 \quad \text{in the operator partial order}, \qquad (2.30)$$

then, by (2.21) we have

$$(1 \le) \frac{\|T\|}{w(T)} \le \frac{M + m}{2\sqrt{mM}}, \qquad (2.31)$$

which is equivalent to

$$(0 \le) \|T\| - w(T) \le \frac{\left(\sqrt{M} - \sqrt{m}\right)^2}{2\sqrt{mM}} w(T), \qquad (2.32)$$

while from (2.24) we have

$$(0 \le) \|T\|^2 - w^2(T) \le \left(\sqrt{M} - \sqrt{m}\right)^2 w(T). \qquad (2.33)$$

Also, the inequality (2.7) becomes

$$(0 \le) \|T\| - w(T) \le \frac{1}{4} \cdot \frac{(M - m)^2}{M + m}. \qquad (2.34)$$

2.2   More Inequalities for Norm and Numerical Radius

2.2.1   A Result via Buzano's Inequality

The following result may be stated as well:

Theorem 36 (Dragomir [16], 2007). *Let $(H; \langle \cdot, \cdot \rangle)$ be a Hilbert space and $T : H \to H$ a bounded linear operator on H. Then*

$$w^2(T) \le \frac{1}{2}\left[w\left(T^2\right) + \|T\|^2\right]. \qquad (2.35)$$

The constant $\frac{1}{2}$ is best possible in (2.35).

Proof. We need the following refinement of Schwarz's inequality obtained by the author in 1985 [4, Theorem 2] (see also [20] and [14]):

$$\|a\| \, \|b\| \geq |\langle a, b \rangle - \langle a, e \rangle \langle e, b \rangle| + |\langle a, e \rangle \langle e, b \rangle| \geq |\langle a, b \rangle|, \tag{2.36}$$

provided a, b, e are vectors in H and $\|e\| = 1$.

Observing that

$$|\langle a, b \rangle - \langle a, e \rangle \langle e, b \rangle| \geq |\langle a, e \rangle \langle e, b \rangle| - |\langle a, b \rangle|,$$

hence by the first inequality in (2.36) we deduce

$$\frac{1}{2} \left(\|a\| \, \|b\| + |\langle a, b \rangle| \right) \geq |\langle a, e \rangle \langle e, b \rangle|. \tag{2.37}$$

This inequality was obtained in a different way earlier by M.L. Buzano in [2].

Now, choose in (2.37), $e = x$, $\|x\| = 1$, $a = Tx$ and $b = T^*x$ to get

$$\frac{1}{2} \left(\|Tx\| \, \|T^*x\| + |\langle T^2 x, x \rangle| \right) \geq |\langle Tx, x \rangle|^2 \tag{2.38}$$

for any $x \in H$, $\|x\| = 1$.

Taking the supremum in (2.38) over $x \in H$, $\|x\| = 1$, we deduce the desired inequality (2.35).

Now, if we assume that (2.35) holds with a constant $C > 0$, *i.e.*,

$$w^2 (T) \leq C \left[w \left(T^2 \right) + \|T\|^2 \right] \tag{2.39}$$

for any $T \in B(H)$, then if we choose T a normal operator and use the fact that for normal operators we have $w(T) = \|T\|$ and $w\left(T^2\right) = \left\|T^2\right\| = \|T\|^2$, then by (2.39) we deduce that $2C \geq 1$ which proves the sharpness of the constant. ∎

Remark 37. From the above result (2.35) we obviously have

$$w(T) \leq \left\{ \frac{1}{2} \left[w \left(T^2 \right) + \|T\|^2 \right] \right\}^{1/2}$$

$$\leq \left\{ \frac{1}{2} \left(\left\|T^2\right\| + \|T\|^2 \right) \right\}^{1/2} \leq \|T\| \tag{2.40}$$

and

$$w(T) \leq \left\{ \frac{1}{2} \left[w \left(T^2 \right) + \|T\|^2 \right] \right\}^{1/2}$$

$$\leq \left\{ \frac{1}{2} \left(w^2 (T) + \|T\|^2 \right) \right\}^{1/2} \leq \|T\|. \tag{2.41}$$

2.2.2 Other Related Results

The following result may be stated.

Theorem 38 (Dragomir [16], 2007). *Let $T : H \to H$ be a bounded linear operator on the Hilbert space H and $\lambda \in \mathbb{C} \setminus \{0\}$. If $\|T\| \leq |\lambda|$, then*

$$\|T\|^{2r} + |\lambda|^{2r} \leq 2 \|T\|^{r-1} |\lambda|^r w(T) + r^2 |\lambda|^{2r-2} \|T - \lambda I\|^2, \qquad (2.42)$$

where $r \geq 1$.

Proof. We use the following inequality for vectors in inner product spaces due to Goldstein, Ryff and Clarke [31]:

$$\|a\|^{2r} + \|b\|^{2r} - 2 \|a\|^r \|b\|^r \frac{\operatorname{Re} \langle a, b \rangle}{\|a\| \|b\|}$$

$$\leq \begin{cases} r^2 \|a\|^{2r-2} \|a - b\|^2 & \text{if } r \geq 1, \\ \\ \|b\|^{2r-2} \|a - b\|^2 & \text{if } r < 1, \end{cases} \qquad (2.43)$$

provided $r \in \mathbb{R}$ and $a, b \in H$ with $\|a\| \geq \|b\|$.

Now, let $x \in H$ with $\|x\| = 1$. From the hypothesis of the theorem, we have that $\|Tx\| \leq |\lambda| \|x\|$ and applying (2.43) for the choices $a = \lambda x$, $\|x\| = 1$, $b = Tx$, we get

$$\|Tx\|^{2r} + |\lambda|^{2r} - 2 \|Tx\|^{r-1} |\lambda|^r |\langle Tx, x \rangle| \leq r^2 |\lambda|^{2r-2} \|Tx - \lambda x\|^2 \qquad (2.44)$$

for any $x \in H$, $\|x\| = 1$ and $r \geq 1$.

Taking the supremum in (2.44) over $x \in H$, $\|x\| = 1$, we deduce the desired inequality (2.42). ∎

The following result may be stated as well:

Theorem 39 (Dragomir [16], 2007). *Let $T : H \to H$ be a bounded linear operator on the Hilbert space $(H, \langle \cdot, \cdot \rangle)$. Then for any $\alpha \in [0, 1]$ and $t \in \mathbb{R}$ one has the inequality:*

$$\|T\|^2 \leq \left[(1 - \alpha)^2 + \alpha^2 \right] w^2(T) + \alpha \|T - tI\|^2$$

$$+ (1 - \alpha) \|T - itI\|^2. \qquad (2.45)$$

Proof. We use the following inequality obtained by the author in [14]:

$$\left[\alpha \|tb - a\|^2 + (1 - \alpha) \|itb - a\|^2 \right] \|b\|^2$$

$$\geq \|a\|^2 \|b\|^2 - [(1 - \alpha) \operatorname{Im} \langle a, b \rangle + \alpha \operatorname{Re} \langle a, b \rangle]^2 \ (\geq 0)$$

to get

$$\|a\|^2 \|b\|^2 \leq [(1 - \alpha) \operatorname{Im} \langle a, b \rangle + \alpha \operatorname{Re} \langle a, b \rangle]^2$$
$$+ \left[\alpha \|tb - a\|^2 + (1 - \alpha) \|itb - a\|^2 \right] \|b\|^2$$
$$\leq \left[(1 - \alpha)^2 + \alpha^2 \right] |\langle a, b \rangle|^2$$
$$+ \left[\alpha \|tb - a\|^2 + (1 - \alpha) \|itb - a\|^2 \right] \|b\|^2 \qquad (2.46)$$

for any $a, b \in H$, $\alpha \in [0, 1]$ and $t \in \mathbb{R}$.

Choosing in (2.46) $a = Tx$, $b = x$, $x \in H$, $\|x\| = 1$, we get

$$\|Tx\|^2 \leq \left[(1 - \alpha)^2 + \alpha^2 \right] |\langle Tx, x \rangle|^2$$
$$+ \alpha \|tx - Tx\|^2 + (1 - \alpha) \|itx - Tx\|^2. \qquad (2.47)$$

Finally, taking the supremum over $x \in H$, $\|x\| = 1$ in (2.47), we deduce the desired result.   ∎

The following particular cases may be of interest.

Corollary 40. *For any T a bounded linear operator on H, one has:*

$$(0 \leq) \|T\|^2 - w^2(T) \leq \begin{cases} \inf\limits_{t \in \mathbb{R}} \|T - tI\|^2 \\[2mm] \inf\limits_{t \in \mathbb{R}} \|T - itI\|^2 \end{cases} \qquad (2.48)$$

and

$$\|T\|^2 \leq \frac{1}{2} w^2(T) + \frac{1}{2} \inf_{t \in \mathbb{R}} \left[\|T - tI\|^2 + \|T - itI\|^2 \right]. \qquad (2.49)$$

Remark 41. The inequality (2.48) can in fact be improved taking into account that for any $a, b \in H$, $b \neq 0$ (see for instance [6]) the bound

$$\inf_{\lambda \in \mathbb{C}} \|a - \lambda b\|^2 = \frac{\|a\|^2 \|b\|^2 - |\langle a, b \rangle|^2}{\|b\|^2}$$

actually implies that

$$\|a\|^2 \|b\|^2 - |\langle a, b \rangle|^2 \leq \|b\|^2 \|a - \lambda b\|^2 \qquad (2.50)$$

for any $a, b \in H$ and $\lambda \in \mathbb{C}$.

Now if in (2.50) we choose $a = Tx$, $b = x$, $x \in H$, $\|x\| = 1$, then we obtain

$$\|Tx\|^2 - |\langle Tx, x \rangle|^2 \leq \|Tx - \lambda x\|^2 \qquad (2.51)$$

for any $\lambda \in \mathbb{C}$, which, by taking the supremum over $x \in H$, $\|x\| = 1$, implies that

$$(0 \leq) \|T\|^2 - w^2(T) \leq \inf_{\lambda \in \mathbb{C}} \|T - \lambda I\|^2. \tag{2.52}$$

Remark 42. If we take $a = x$, $b = Tx$ in (2.50), then we obtain

$$\|Tx\|^2 \leq |\langle Tx, x \rangle|^2 + \|Tx\|^2 \|x - \mu Tx\|^2 \tag{2.53}$$

for any $x \in H$, $\|x\| = 1$ and $\mu \in \mathbb{C}$. Now, if we take the supremum over $x \in H$, $\|x\| = 1$ in (2.53), then we get

$$(0 \leq) \|T\|^2 - w^2(T) \leq \|T\|^2 \inf_{\mu \in \mathbb{C}} \|I - \mu T\|^2. \tag{2.54}$$

Finally and from a different view point we may state:

Theorem 43 (Dragomir [16], 2007). *Let* $T : H \to H$ *be a bounded linear operator on* H. *If* $p \geq 2$, *then:*

$$\|T\|^p + |\lambda|^p \leq \frac{1}{2} \left(\|T + \lambda I\|^p + \|T - \lambda I\|^p \right), \tag{2.55}$$

for any $\lambda \in \mathbb{C}$.

Proof. We use the following inequality obtained by Dragomir and Sándor in [20]:

$$\|a + b\|^p + \|a - b\|^p \geq 2 \left(\|a\|^p + \|b\|^p \right) \tag{2.56}$$

for any $a, b \in H$ and $p \geq 2$.

Now, if we choose $a = Tx$, $b = \lambda x$, then we get

$$\|Tx + \lambda x\|^p + \|Tx - \lambda x\|^p \geq 2 \left(\|Tx\|^p + |\lambda|^p \right) \tag{2.57}$$

for any $x \in H$, $\|x\| = 1$.

Taking the supremum in (2.57) over $x \in H$, $\|x\| = 1$, we get the desired result (2.55). ∎

Remark 44. For $p = 2$, we have the simpler result:

$$\|T\|^2 + |\lambda|^2 \leq \frac{1}{2} \left(\|T + \lambda I\|^2 + \|T - \lambda I\|^2 \right) \tag{2.58}$$

for any $\lambda \in \mathbb{C}$. This can easily be obtained from the parallelogram identity as well.

2.3　Some Associated Functionals

2.3.1　Some Fundamental Facts

Replacing the supremum with the infimum in the definitions of the operator norm and numerical radius, we can also consider the quantities $\ell(T) := \inf_{\|x\|=1} \|Tx\|$ and $m(T) = \inf_{\|x\|=1} |\langle Tx, x \rangle|$. By the *Schwarz inequality*, it is obvious that $m(T) \le \ell(T)$ for each $T \in B(H)$.

We can also consider the functionals $v_s, \delta_s : B(H) \to \mathbb{R}$ introduced in [17] and given by

$$v_s(T) := \sup_{\|x\|=1} \operatorname{Re} \langle Tx, x \rangle \quad \text{and} \quad \delta_s(T) := \sup_{\|x\|=1} \operatorname{Im} \langle Tx, x \rangle \qquad (2.59)$$

where "s" stands for *supremum*, while the corresponding ones for *infimum* are defined as:

$$v_i(T) := \inf_{\|x\|=1} \operatorname{Re} \langle Tx, x \rangle \quad \text{and} \quad \delta_i(T) := \inf_{\|x\|=1} \operatorname{Im} \langle Tx, x \rangle. \qquad (2.60)$$

We notice that the functionals v_p, δ_p with $p \in \{s, i\}$ are obviously connected by the formula

$$\delta_p(T) = -v_q(iT) \quad \text{for any} \quad T \in B(H), \qquad (2.61)$$

where $p \ne q$ and the "i" in front of T represents the imaginary unit. Also, by definition, v_s and δ_s are *positive homogeneous* and *subadditive* while v_i and δ_i are positive homogeneous and *superadditive*.

Due to the fact that for any $x \in H$, $\|x\| = 1$ we have

$$-w(T) \le -|\langle Tx, x \rangle| \le \operatorname{Re} \langle Tx, x \rangle,$$
$$\operatorname{Im} \langle Tx, x \rangle \le |\operatorname{Im} \langle Tx, x \rangle| \le w(T),$$

then, by taking the supremum and the infimum, respectively, over $x \in H$, $\|x\| = 1$, we deduce the simple inequality:

$$\max \left\{ |v_p(T)|, |\delta_p(T)| \right\} \le w(T), \quad T \in B(H) \qquad (2.62)$$

where $p \in \{s, i\}$.

The main aim of this section is twofold. Firstly, some natural connections amongst the functionals v_p, δ_p, the operator norm and the numerical ranges w, m, w_e and m_e are pointed out. Secondly, some new inequalities for operators $T \in B(H)$ for which the composite operator $C_{\gamma, \Gamma}(T)$ with $\gamma, \Gamma \in \mathbb{K}$ is assumed to be c^2-accretive with $c \in \mathbb{R}$ are also given. New upper bounds for the nonnegative quantity $\|T\|^2 - w^2(T)$ are obtained as well.

2.3.2 Preliminary Results

In the following we establish an identity connecting the numerical radius of an operator with the other functionals defined in the introduction of this section.

Lemma 45. *Let $T \in B(H)$ and $\gamma, \Gamma \in \mathbb{K}$. Then for any $x \in H$, $\|x\| = 1$ we have the equality:*

$$\mathrm{Re}\left[\langle(\Gamma I - T)x, x\rangle \langle x, (T - \gamma I)x\rangle\right]$$

$$= \frac{1}{4}|\Gamma - \gamma|^2 - \left|\left\langle\left(T - \frac{\gamma + \Gamma}{2} \cdot I\right)x, x\right\rangle\right|^2. \tag{2.63}$$

Proof. We use the following elementary identity for complex numbers:

$$\mathrm{Re}\left(a\bar{b}\right) = \frac{1}{4}\left[|a + b|^2 - |a - b|^2\right], \quad a, b \in \mathbb{C}, \tag{2.64}$$

for the choices $a = \langle(\Gamma I - T)x, x\rangle = \Gamma - \langle Tx, x\rangle$ and $b = \langle(T - \gamma I)x, x\rangle = \langle Tx, x\rangle - \gamma$ to get

$$\mathrm{Re}\left[\langle(\Gamma I - T)x, x\rangle \overline{\langle(T - \gamma I)x, x\rangle}\right]$$

$$= \frac{1}{4}\left[|\Gamma - \gamma|^2 - |\langle 2\langle Tx, x\rangle - (\gamma + \Gamma)\rangle|^2\right] \tag{2.65}$$

for $x \in H$, $\|x\| = 1$, which is clearly equivalent with (2.63). ∎

Corollary 46. *For any $T \in B(H)$ and $\gamma, \Gamma \in \mathbb{K}$ we have*

$$\inf_{\|x\|=1} \mathrm{Re}\left[\langle(\Gamma I - T)x, x\rangle \langle x, (T - \gamma I)x\rangle\right]$$

$$= \frac{1}{4}|\Gamma - \gamma|^2 - w^2\left(T - \frac{\gamma + \Gamma}{2} \cdot I\right) \tag{2.66}$$

and

$$\sup_{\|x\|=1} \mathrm{Re}\left[\langle(\Gamma I - T)x, x\rangle \langle x, (T - \gamma I)x\rangle\right]$$

$$= \frac{1}{4}|\Gamma - \gamma|^2 - m^2\left(T - \frac{\gamma + \Gamma}{2} \cdot I\right). \tag{2.67}$$

The proof is obvious from the identity (2.63) on taking the infimum and the supremum over $x \in H$, $\|x\| = 1$, respectively.

If we denote by $S_H := \{x \in H \mid \|x\| = 1\}$ the unit sphere in H and, for $T \in B(H)$, $\gamma, \Gamma \in \mathbb{K}$ we define [17]

$$\mu(T; \gamma, \Gamma)(x) := \mathrm{Re}\left[\langle(\Gamma I - T)x, x\rangle \langle x, (T - \gamma I)x\rangle\right], \quad x \in S_H;$$

then, on utilizing the elementary properties of complex numbers, we have

$$\mu\left(T;\gamma,\Gamma\right)(x) = \left(\operatorname{Re}\Gamma - \operatorname{Re}\langle Tx, x\rangle\right)\left(\operatorname{Re}\langle Tx, x\rangle - \operatorname{Re}\gamma\right)$$

$$+ \left(\operatorname{Im}\Gamma - \operatorname{Im}\langle Tx, x\rangle\right)\left(\operatorname{Im}\langle Tx, x\rangle - \operatorname{Im}\gamma\right) \quad (2.68)$$

for any $x \in S_H$.

If we denote [17]:

$$\mu_{s(i)}\left(T;\gamma,\Gamma\right) := \sup_{\|x\|=1}\left(\inf_{\|x\|=1}\right)\mu\left(T;\gamma,\Gamma\right)(x)$$

then (2.66) can be stated as:

$$\mu_i\left(T;\gamma,\Gamma\right) + w^2\left(T - \frac{\gamma + \Gamma}{2} \cdot I\right) = \frac{1}{4}\left|\Gamma - \gamma\right|^2 \quad (2.69)$$

while (2.67) can be stated as:

$$\mu_s\left(T;\gamma,\Gamma\right) + m^2\left(T - \frac{\gamma + \Gamma}{2} \cdot I\right) = \frac{1}{4}\left|\Gamma - \gamma\right|^2 \quad (2.70)$$

for any $T \in B\left(H\right)$ and $\gamma, \Gamma \in \mathbb{K}$.

Remark 47. Utilizing the equality (2.68), a sufficient condition for the inequality $\mu_i\left(T;\gamma,\Gamma\right) \geq 0$ or, equivalently, $w\left(T - \frac{\gamma+\Gamma}{2} \cdot I\right) \leq \frac{1}{2}\left|\Gamma - \gamma\right|$ to hold is that

$$\operatorname{Re}\Gamma \geq \operatorname{Re}\langle Tx, x\rangle \geq \operatorname{Re}\gamma \quad \text{and} \quad \operatorname{Im}\Gamma \geq \operatorname{Im}\langle Tx, x\rangle \geq \operatorname{Im}\gamma \quad (2.71)$$

for each $x \in H, \|x\| = 1$.

The following identity that links the norm with the inner product also holds.

Lemma 48. *Let $T \in B\left(H\right)$ and $\gamma, \Gamma \in \mathbb{K}$. The for each $x \in H, \|x\| = 1$, we have the equality:*

$$\operatorname{Re}\langle\left(T^* - \bar{\gamma}I\right)\left(\Gamma I - T\right)x, x\rangle = \frac{1}{4}\left|\Gamma - \gamma\right|^2 - \left\|\left(T - \frac{\gamma + \Gamma}{2} \cdot I\right)x\right\|^2. \quad (2.72)$$

Proof. We utilize the simple identity in inner product spaces

$$\operatorname{Re}\langle u - y, y - v\rangle = \frac{1}{4}\|u - v\|^2 - \left\|y - \frac{u + v}{2}\right\|^2, \quad (2.73)$$

$u, v, y \in H$, for the choices $u = \Gamma x, y = Tx, v = \gamma x$ with $x \in H, \|x\| = 1$ to get

$$\operatorname{Re}\langle\Gamma x - Tx, Tx - \gamma x\rangle = \frac{1}{4}\left|\Gamma - \gamma\right|^2 - \left\|\left(T - \frac{\gamma + \Gamma}{2} \cdot I\right)x\right\|^2,$$

$x \in H, \|x\| = 1$, which is clearly equivalent with (2.72). ∎

Corollary 49. *For any $T \in B(H)$ and $\gamma, \Gamma \in \mathbb{K}$ we have*

$$v_i \left[(T^* - \bar{\gamma} I)(\Gamma I - T) \right]$$
$$= \frac{1}{4} |\Gamma - \gamma|^2 - \left\| T - \frac{\gamma + \Gamma}{2} \cdot I \right\|^2 \tag{2.74}$$

and

$$v_s \left[(T^* - \bar{\gamma} I)(\Gamma I - T) \right]$$
$$= \frac{1}{4} |\Gamma - \gamma|^2 - \ell^2 \left(T - \frac{\gamma + \Gamma}{2} \cdot I \right). \tag{2.75}$$

We recall that a bounded linear operator $T : H \rightarrow H$ is called *strongly c^2-accretive* (with $c \neq 0$) if $\operatorname{Re} \langle Ty, y \rangle \geq c^2$ for each $y \in H$, $\|y\| = 1$. For $c = 0$, the operator is called *accretive*. Therefore, and for the sake of simplicity, we can call the operator c^2-*accretive* for $c \in \mathbb{R}$ and understand the statement in the above sense.

Utilizing the identity (2.72) we can state the following characterization result that will be useful in the sequel:

Lemma 50 (Dragomir [17], 2007). *For $T \in B(H)$ and $\gamma, \Gamma \in \mathbb{K}$, $c \in \mathbb{R}$, the following statements are equivalent:*

(i) *The operator $C_{\gamma, \Gamma}(T) := (T^* - \bar{\gamma} I)(\Gamma I - T)$ is c^2-accretive;*
(ii) *We have the inequality:*

$$\left\| T - \frac{\gamma + \Gamma}{2} \cdot I \right\|^2 \leq \frac{1}{4} |\Gamma - \gamma|^2 - c^2. \tag{2.76}$$

Remark 51. Since the self-adjoint operator $T : H \rightarrow H$ satisfying the condition: $T \geq c^2 I$ in the operator partial order "$\geq$" is c^2-accretive, then a sufficient condition for $C_{\gamma, \Gamma}(T)$ to be c^2-accretive is that $C_{\gamma, \Gamma}(T)$ is self-adjoint and $C_{\gamma, \Gamma}(T) \geq c^2 I$.

2.3.3 General Inequalities

We can state the following result that provides some inequalities between different numerical radii:

Theorem 52 (Dragomir [17], 2007). *For any $T \in B(H)$ and $\gamma, \Gamma \in \mathbb{K}$ we have the inequalities*

$$\frac{1}{4} |\Gamma - \gamma|^2 \leq m^2 \left(T - \frac{\gamma + \Gamma}{2} \cdot I \right)$$
$$+ \begin{cases} \frac{1}{2} w_e^2 (\Gamma I - T, T - \gamma I) \\ w (\Gamma I - T) w (T - \gamma I) \end{cases} \tag{2.77}$$

and

$$\frac{1}{4}|\Gamma - \gamma|^2 \le w^2 \left(T - \frac{\gamma + \Gamma}{2} \cdot I \right) + \frac{1}{2} m_e^2 \left(\Gamma I - T, T - \gamma I \right). \qquad (2.78)$$

Proof. Utilizing the elementary inequality

$$\text{Re} \left(a\bar{b} \right) \le \frac{1}{2} \left[|a|^2 + |b|^2 \right], \quad a, b \in \mathbb{C} \qquad (2.79)$$

we can state that

$$\text{Re} \left[\langle (\Gamma I - T) x, x \rangle \overline{\langle (T - \gamma I) x, x \rangle} \right]$$

$$\le \frac{1}{2} \left[|\langle (\Gamma I - T) x, x \rangle|^2 + |\langle (T - \gamma I) x, x \rangle|^2 \right] \qquad (2.80)$$

for any $x \in H$, $\|x\| = 1$.

Taking the supremum over $x \in H$, $\|x\| = 1$ in (2.80) and utilizing the representation (2.67) in Corollary 46, we deduce

$$\frac{1}{4}|\Gamma - \gamma|^2 - m^2 \left(T - \frac{\gamma + \Gamma}{2} \cdot I \right)$$

$$\le \frac{1}{2} \sup_{\|x\|=1} \left[|\langle (\Gamma I - T) x, x \rangle|^2 + |\langle (T - \gamma I) x, x \rangle|^2 \right]$$

$$= \frac{1}{2} w_e^2 \left(\Gamma I - T, T - \gamma I \right),$$

which is clearly equivalent to the first inequality in (2.77).

Now, by the elementary inequality

$$\text{Re} \left(a\bar{b} \right) \le |a| |b| \quad \text{for each } a, b \in \mathbb{C},$$

we can also state that

$$\frac{1}{4}|\Gamma - \gamma|^2 - m^2 \left(T - \frac{\gamma + \Gamma}{2} \cdot I \right)$$

$$\le \sup_{\|x\|=1} \left[|\langle (T - \Gamma I) x, x \rangle| \, |\langle (T - \gamma I) x, x \rangle| \right]$$

$$\le \sup_{\|x\|=1} |\langle (T - \Gamma I) x, x \rangle| \cdot \sup_{\|x\|=1} |\langle (T - \gamma I) x, x \rangle|$$

$$= w \left(\Gamma I - T \right) w \left(T - \gamma I \right)$$

and the second part of (2.77) is also proved.

Taking the infimum over $x \in H$, $\|x\| = 1$ in (2.80) and making use of the representation (2.66) from Corollary 46, we deduce the inequality in (2.78). ∎

Remark 53. If the operator $T \in B(H)$ and the complex numbers γ, Γ are such that $\mu_i(T; \gamma, \Gamma) \geq 0$ or, equivalently $w\left(T - \frac{\gamma + \Gamma}{2} I\right) \leq \frac{1}{2}|\Gamma - \gamma|$, then we have the reverse inequalities

$$0 \leq \frac{1}{4}|\Gamma - \gamma|^2 - m^2\left(T - \frac{\gamma + \Gamma}{2} \cdot I\right)$$

$$\leq \begin{cases} \frac{1}{2}w_e^2\left(\Gamma I - T, T - \gamma I\right) \\ w\left(\Gamma I - T\right) w\left(T - \gamma I\right) \end{cases} \tag{2.81}$$

and

$$0 \leq \frac{1}{4}|\Gamma - \gamma|^2 - w^2\left(T - \frac{\gamma + \Gamma}{2} \cdot I\right)$$

$$\leq \frac{1}{2}m_e^2\left(\Gamma I - T, T - \gamma I\right). \tag{2.82}$$

Since, in general, $w(B) \leq \|B\|$, $B \in B(H)$, hence a sufficient condition for (2.81) and (2.82) to hold is that $\left\|T - \frac{\gamma + \Gamma}{2} I\right\| \leq \frac{1}{2}|\Gamma - \gamma|$ holds true. We also notice that this last condition is equivalent with the fact that the operator $C_{\gamma,\Gamma}(T) = (T^* - \bar{\gamma}I)(\Gamma I - T)$ is accretive.

From a different perspective and as pointed out in Remark 47, a sufficient condition for $\mu_i(T; \gamma, \Gamma) \geq 0$ to hold is that (2.71) holds true and, therefore, if (2.71) is valid, then both (2.81) and (2.82) can be stated.

The following reverse inequality of (2.82) is incorporated in the following result:

Proposition 54 (Dragomir [17], 2007). *Let $T \in B(H)$ and $\gamma, \Gamma \in \mathbb{K}$ be such that (2.71) holds true. Then*

$$(0 \leq) \left(\operatorname{Re}\Gamma - \nu_s(T)\right)\left(\nu_i(T) - \operatorname{Re}\gamma\right)$$

$$+ \left(\operatorname{Im}\Gamma - \delta_s(T)\right)\left(\delta_i(T) - \operatorname{Im}\gamma\right)$$

$$\leq \frac{1}{4}|\Gamma - \gamma|^2 - w^2\left(T - \frac{\gamma + \Gamma}{2} \cdot I\right). \tag{2.83}$$

Proof. Taking the infimum for $x \in H$, $\|x\| = 1$ in the identity (2.68) and utilizing the representation (2.66) and the properties of infimum, we have

$$\frac{1}{4}|\Gamma - \gamma|^2 - w^2\left(T - \frac{\gamma + \Gamma}{2} \cdot I\right)$$

$$\geq \inf_{\|x\|=1} \left[\left(\operatorname{Re}\Gamma - \operatorname{Re}\langle Tx, x\rangle\right)\left(\operatorname{Re}\langle Tx, x\rangle - \operatorname{Re}\gamma\right)\right]$$

$$+ \inf_{\|x\|=1} \left[\left(\operatorname{Im}\Gamma - \operatorname{Im}\langle Tx, x\rangle\right)\left(\operatorname{Im}\langle Tx, x\rangle - \operatorname{Im}\gamma\right)\right]$$

$$\geq \inf_{\|x\|=1} \left(\operatorname{Re} \Gamma - \operatorname{Re} \langle Tx, x \rangle \right) \cdot \inf_{\|x\|=1} \left(\operatorname{Re} \langle Tx, x \rangle - \operatorname{Re} \gamma \right)$$

$$+ \inf_{\|x\|=1} \left(\operatorname{Im} \Gamma - \operatorname{Im} \langle Tx, x \rangle \right) \cdot \inf_{\|x\|=1} \left(\operatorname{Im} \langle Tx, x \rangle - \operatorname{Im} \gamma \right)$$

$$= \left(\operatorname{Re} \Gamma - \sup_{\|x\|=1} \operatorname{Re} \langle Tx, x \rangle \right) \left(\inf_{\|x\|=1} \operatorname{Re} \langle Tx, x \rangle - \operatorname{Re} \gamma \right)$$

$$+ \left(\operatorname{Im} \Gamma - \sup_{\|x\|=1} \operatorname{Im} \langle Tx, x \rangle \right) \left(\inf_{\|x\|=1} \operatorname{Im} \langle Tx, x \rangle - \operatorname{Im} \gamma \right)$$

which is exactly the desired result (2.83).   ∎

The representation in Lemma 48 has its natural consequences relating the numerical values $\ell(T)$ and $w(T)$ of certain operators as described in the following:

Theorem 55 (Dragomir [17], 2007). *For any $T \in B(H)$ and $\gamma, \Gamma \in \mathbb{K}$ we have*

$$\frac{1}{4} |\Gamma - \gamma|^2 \leq \ell^2 \left(T - \frac{\gamma + \Gamma}{2} \cdot I \right)$$

$$+ \begin{cases} \frac{1}{2} w \left[(\bar{\Gamma} I - T^*) (\Gamma I - T) + (T^* - \bar{\gamma} I) (T - \gamma I) \right], \\[2mm] w \left[(T^* - \bar{\gamma} I) (\Gamma I - T) \right], \\[2mm] \frac{1}{4} \left\| (T^* - \bar{\gamma} I) (\Gamma I - T) - I \right\|^2 \end{cases} \qquad (2.84)$$

and

$$\frac{1}{4} |\Gamma - \gamma|^2 \leq \left\| T - \frac{\gamma + \Gamma}{2} I \right\|^2$$

$$+ \begin{cases} \frac{1}{2} m \left[(\bar{\Gamma} I - T^*) (\Gamma I - T) + (T^* - \bar{\gamma} I) (T - \gamma I) \right], \\[2mm] m \left[(T^* - \bar{\gamma} I) (\Gamma I - T) \right], \\[2mm] \frac{1}{4} \ell^2 \left[(T^* - \bar{\gamma} I) (\Gamma I - T) - I \right], \end{cases} \qquad (2.85)$$

respectively.

Proof. Utilizing the elementary inequality in inner product spaces

$$\operatorname{Re} \langle u, v \rangle \leq \frac{1}{2} \left[\|u\|^2 + \|v\|^2 \right], \qquad u, v \in H, \qquad (2.86)$$

we can state that

$$\operatorname{Re} \langle (\Gamma I - T) x, (T - \gamma I) x \rangle$$

$$\leq \frac{1}{2} \left[\|(\Gamma I - T) x\|^2 + \|(T - \gamma I) x\|^2 \right]$$

$$= \frac{1}{2} \left[\left\langle \left(\bar{\Gamma} I - T^* \right) \left(\Gamma I - T \right) x, x \right\rangle + \left\langle \left(T^* - \bar{\gamma} I \right) \left(T - \gamma I \right) x, x \right\rangle \right]$$

$$= \frac{1}{2} \left\langle \left[\left(\bar{\Gamma} I - T^* \right) \left(\Gamma I - T \right) + \left(T^* - \bar{\gamma} I \right) \left(T - \gamma I \right) \right] x, x \right\rangle \qquad (2.87)$$

for each $x \in H$, $\|x\| = 1$.

Taking the supremum in (2.87) over $x \in H$, $\|x\| = 1$ and utilizing the representation (2.75), we deduce the first inequality in (2.84).

Now, by the elementary inequality $\mathrm{Re}\,(T) \leq |T|$, $T \in \mathbb{C}$ we have

$$\mathrm{Re} \left\langle \left(T^* - \bar{\gamma} I \right) \left(\Gamma I - T \right) x, x \right\rangle \leq \left| \left\langle \left(T^* - \bar{\gamma} I \right) \left(\Gamma I - T \right) x, x \right\rangle \right|, \qquad (2.88)$$

which provides, by taking the supremum over $x \in H$, $\|x\| = 1$, the second inequality in (2.84).

Finally, on utilizing the inequality

$$\mathrm{Re} \, \langle u, v \rangle \leq \frac{1}{4} \|u - v\|^2, \qquad u, v \in H,$$

we also have

$$\mathrm{Re} \left\langle \left(T^* - \bar{\gamma} I \right) \left(\Gamma I - T \right) x, x \right\rangle \leq \frac{1}{4} \left\| \left[\left(T^* - \bar{\gamma} I \right) \left(\Gamma I - T \right) - I \right] x \right\|^2 \qquad (2.89)$$

for any $x \in H$, $\|x\| = 1$, which gives, by taking the supremum, the last part of (2.84).

The proof of (2.85) follows by the representation (2.74) in Corollary 49 and by the inequalities (2.87)–(2.89) above in which we take the infimum over $x \in H$, $\|x\| = 1$. ∎

Corollary 56. *Let* $T \in B(H)$ *and* $\gamma, \Gamma \in \mathbb{K}$. *If* $C_{\gamma, \Gamma}(T)$ *is accretive, then*

$$0 \leq \frac{1}{4} |\Gamma - \gamma|^2 - \ell^2 \left(T - \frac{\gamma + \Gamma}{2} \cdot I \right)$$

$$\leq \begin{cases} \frac{1}{2} w \left[\left(\bar{\Gamma} I - T^* \right) \left(\Gamma I - T \right) + \left(T^* - \bar{\gamma} I \right) \left(T - \gamma I \right) \right], \\ w \left[\left(T^* - \bar{\gamma} I \right) \left(\Gamma I - T \right) \right], \\ \frac{1}{4} \left\| \left(T^* - \bar{\gamma} I \right) \left(\Gamma I - T \right) - I \right\|^2 \end{cases} \qquad (2.90)$$

and

$$0 \leq \frac{1}{4} |\Gamma - \gamma|^2 - \left\| T - \frac{\gamma + \Gamma}{2} I \right\|^2$$

$$\leq \begin{cases} \frac{1}{2}m\left[\left(\bar{\Gamma}I - T^*\right)\left(\Gamma I - T\right) + \left(T^* - \bar{\gamma}I\right)\left(T - \gamma I\right)\right], \\ m\left[\left(T^* - \bar{\gamma}I\right)\left(\Gamma I - T\right)\right], \\ \frac{1}{4}\ell^2\left[\left(T^* - \bar{\gamma}I\right)\left(\Gamma I - T\right) - I\right], \end{cases} \quad (2.91)$$

respectively.

2.3.4   Reverse Inequalities

The inequality $\|T\| \geq w(T)$ for any bounded linear operator $T \in B(H)$ is a fundamental result in Operator Theory and therefore it is useful to know some upper bounds for the nonnegative quantity $\|T\| - w(T)$ under various assumptions for the operator T. In our recent paper [13] several such inequalities have been obtained. In order to establish some new results that would complement the inequalities outlined in the Introduction, we need the following lemma which provides two simple identities of interest:

Lemma 57 (Dragomir [17], 2007). *For any $T \in B(H)$ and $\gamma, \Gamma \in \mathbb{K}$ we have*

$$\|Tx\|^2 - |\langle Tx, x\rangle|^2$$

$$= \left\|\left(T - \frac{\gamma + \Gamma}{2} \cdot I\right)x\right\|^2 - \left|\left\langle\left(T - \frac{\gamma + \Gamma}{2} \cdot I\right)x, x\right\rangle\right|^2$$

$$= \operatorname{Re}\left[\langle(\Gamma I - T)x, x\rangle\langle x, (T - \gamma I)x\rangle\right] - \operatorname{Re}\langle(\Gamma I - T)x, (T - \gamma I)x\rangle, \quad (2.92)$$

for each $x \in H$, $\|x\| = 1$.

Proof. The first identity is obvious by direct calculation. The second identity can be obtained, for instance, by subtracting the identity (2.72) from (2.63).   ∎

As a natural application of the above lemma in providing upper bounds for the nonnegative quantity $\|T\|^2 - w^2(T)$, $T \in B(H)$, we can state the following result:

Theorem 58 (Dragomir [17], 2007). *For any $T \in B(H)$ and $\gamma, \Gamma \in \mathbb{K}$ we have*

$$(0 \leq) \|T\|^2 - w^2(T)$$

$$\leq \left\|T - \frac{\gamma + \Gamma}{2}I\right\|^2 - m^2\left(T - \frac{\gamma + \Gamma}{2} \cdot I\right)$$

$$= \frac{1}{4}|\Gamma - \gamma|^2 - m^2\left(T - \frac{\gamma + \Gamma}{2} \cdot I\right) - v_i\left[\left(T^* - \bar{\gamma}I\right)\left(\Gamma I - T\right)\right]. \quad (2.93)$$

Proof. From the first identity in (2.92) we have

$$\|Tx\|^2 = |\langle Tx, x \rangle|^2 + \left\| \left(T - \frac{\gamma + \Gamma}{2} \cdot I \right) x \right\|^2 \tag{2.94}$$

$$- \left| \left\langle \left(T - \frac{\gamma + \Gamma}{2} \cdot I \right) x, x \right\rangle \right|^2$$

for any $x \in H$, $\|x\| = 1$.

Taking the supremum over $x \in H$, $\|x\| = 1$ and utilizing the fact that

$$\sup_{\|x\|=1} \left[|\langle Tx, x \rangle|^2 + \left\| \left(T - \frac{\gamma + \Gamma}{2} \cdot I \right) x \right\|^2 - \left| \left\langle \left(T - \frac{\gamma + \Gamma}{2} \cdot I \right) x, x \right\rangle \right|^2 \right]$$

$$\leq \sup_{\|x\|=1} |\langle Tx, x \rangle|^2 + \sup_{\|x\|=1} \left\| \left(T - \frac{\gamma + \Gamma}{2} \cdot I \right) x \right\|^2$$

$$- \inf_{\|x\|=1} \left| \left\langle \left(T - \frac{\gamma + \Gamma}{2} \cdot I \right) x, x \right\rangle \right|^2$$

$$= w^2 (T) + \left\| \left(T - \frac{\gamma + \Gamma}{2} \cdot I \right) x \right\|^2 - m^2 \left(T - \frac{\gamma + \Gamma}{2} \cdot I \right),$$

we deduce the first part of (2.93).

The second part follows by the identity (2.74). ∎

Remark 59. Utilizing the inequality (2.77) in Theorem 52 we can obtain from (2.93) the following result:

$$(0 \leq) \|T\|^2 - w^2 (T)$$

$$\leq -v_i \left[(T^* - \bar{\gamma} I) (\Gamma I - T) \right] + \begin{cases} \frac{1}{2} w_e^2 (\Gamma I - T, T - \gamma I), \\ w (\Gamma I - T) w (T - \gamma I), \end{cases} \tag{2.95}$$

which holds true for each $T \in B(H)$ and $\gamma, \Gamma \in \mathbb{K}$.

Since $m^2 \left(T - \frac{\gamma + \Gamma}{2} I \right) \geq 0$, hence we also have the general inequality

$$(0 \leq) \|T\|^2 - w^2 (T)$$

$$\leq \frac{1}{4} |\Gamma - \gamma|^2 - v_i \left[(T^* - \bar{\gamma} I) (\Gamma I - T) \right], \tag{2.96}$$

for any $T \in B(H)$ and $\gamma, \Gamma \in \mathbb{K}$.

Theorem 58 admits the following particular case that provides a simple reverse inequality for $\|T\| \geq w(T)$ under some appropriate assumptions for the operator T that have been considered in the introduction and are motivated by earlier results:

Corollary 60. *Let $T \in B(H)$ and $\gamma, \Gamma \in \mathbb{K}$, $c \in \mathbb{R}$. If the composite operator $C_{\gamma,\Gamma}(T)$ is c^2-accretive, then:*

$$(0 \leq) \|T\|^2 - w^2(T)$$

$$\leq \frac{1}{4}|\Gamma - \gamma|^2 - c^2 - m^2\left(T - \frac{\gamma + \Gamma}{2} \cdot I\right). \qquad (2.97)$$

The proof is obvious by the first part of the inequality (2.93) and by Lemma 50 which states that $C_{\gamma,\Gamma}(T)$ is c^2-accretive if and only if the inequality (2.73) holds true.

Remark 61. From (2.97) we can deduce the following reverse inequalities which are coarser, but perhaps more useful when the terms in the upper bounds are known:

$$(0 \leq) \|T\|^2 - w^2(T)$$

$$\leq -c^2 + \begin{cases} \frac{1}{4}|\Gamma - \gamma|^2, \\ \frac{1}{2}w_e^2(\Gamma I - T, T - \gamma I), \\ w(\Gamma I - T)w(T - \gamma I). \end{cases} \qquad (2.98)$$

In particular, if $C_{\gamma,\Gamma}(T)$ is accretive, then the following inequalities can be stated:

$$(0 \leq) \|T\|^2 - w^2(T)$$

$$\leq \frac{1}{4}|\Gamma - \gamma|^2 - m^2\left(T - \frac{\gamma + \Gamma}{2} \cdot I\right)$$

$$\leq \begin{cases} \frac{1}{4}|\Gamma - \gamma|^2, \\ \frac{1}{2}w_e^2(\Gamma I - T, T - \gamma I), \\ w(\Gamma I - T)w(T - \gamma I). \end{cases} \qquad (2.99)$$

Remark 62. If $N \geq n > 0$ and the composite operator $C_{n,N}(T)$ is c^2-accretive or, sufficiently, self-adjoint and positive definite with the constant $c^2 \geq 0$, then we have the inequalities:

$$(0 \leq) \|T\|^2 - w^2(T)$$

$$\leq \frac{1}{4}(N - n)^2 - c^2 - m^2\left(T - \frac{\gamma + \Gamma}{2} \cdot I\right)$$

$$\leq -c^2 + \begin{cases} \frac{1}{4}(N-n)^2, \\ \frac{1}{2}w_e^2(NI-T, T-nI), \\ w(NI-T)w(T-nI). \end{cases} \tag{2.100}$$

Remark 63. If the operator T on the scalars γ, Γ from the statement of Corollary 60 have in addition the property that

$$\left| \left\langle \left(T - \frac{\gamma+\Gamma}{2}\cdot I\right)x, x\right\rangle\right| \geq d \quad \text{for each } x \in H, \ \|x\| = 1, \tag{2.101}$$

where $d > 0$ is given, then by (2.97) we also have

$$(0 \leq) \|T\|^2 - w^2(T) \leq \frac{1}{4}|\Gamma - \gamma|^2 - c^2 - d^2. \tag{2.102}$$

We notice that a sufficient condition for (2.101) to hold is that the operator $T - \frac{\gamma+\Gamma}{2}\cdot I$ be d-accretive.

Remark 64. Finally, we note that if the operator $C_{n,N}(T)$ is accretive (or sufficiently self-adjoint and positive), then we have the following reverse inequalities:

$$(0 \leq) \|T\|^2 - w^2(T) \leq \begin{cases} \frac{1}{4}(N-n)^2, \\ \frac{1}{2}w_e^2(NI-T, T-nI), \\ w(NI-T)w(T-nI). \end{cases} \tag{2.103}$$

2.4 Inequalities for the Maximum of the Real Part

2.4.1 Introduction

For a bounded linear operator T on the complex Hilbert space, consider *the maximum and the minimum of the spectrum of the real part of T* denoted by [18]

$$v_{s(i)}(T) := \sup_{\|x\|=1}\left(\inf_{\|x\|=1}\right) \text{Re}\langle Tx, x\rangle = \lambda_{\max(\min)}(\text{Re}(T)).$$

The following properties are obvious by the definition:

  (a) $v_s(-T) = -v_i(T), T \in B(H)$;
 (aa) $v_i(T) \geq 0$ for accretive operators on H;
(aaa) $v_{s(i)}(A+B) \leq (\geq) v_{s(i)}(A) + v_{s(i)}(B)$ for any $A, B \in B(H)$;
 (av) $\max\{|v_i(A)|, |v_s(A)|\} = w(\text{Re}(A)) \leq w(A)$ for all $A \in B(H)$.

More properties which connect these functionals with the semi-inner products generated by the operator norm and the numerical radius are outlined in the next section. An improvement of Lumer's classical result and some bounds is also given.

Motivated by the above results, we establish in the present section some upper bounds for the nonnegative quantities $\|A\| - v_s(\mu A) (\geq \|A\| - w(A) \geq 0)$ and $w(A) - v_s(\mu A) (\geq 0)$, for some $\mu \in \mathbb{C}$, $|\mu| = 1$ under suitable assumptions on the involved operator $A \in B(H)$. Lower bounds for the quantities $\frac{v_s(\mu A)}{\|A\|} \left(\leq \frac{w(A)}{\|A\|} \leq 1 \right)$ and $\frac{v_s(\mu A)}{w(A)} (\leq 1)$ are also given. They improve some results from the earlier paper [13]. Inequalities in terms of the semi-inner products that can naturally be associated with the operator norm and the numerical radius are provided as well.

For other recent results concerning inequalities between the operator norm and numerical radius see the papers [12, 13, 16, 39] and [38]. Lower bounds for $w(A)$ are in the finite-dimensional case studied in [43]. For classical results, see the books [33, 34] and the references therein.

2.4.2 Preliminary Results for Semi-Inner Products

In any normed linear space $(E, \|\cdot\|)$, since the function $f : E \to \mathbb{R}$, $f(x) = \frac{1}{2} \|x\|^2$ is convex, one can introduce the following semi-inner products (see for instance [10]):

$$\langle x, y \rangle_i := \lim_{t \to 0-} \frac{\|y + tx\|^2 - \|y\|^2}{2t}, \tag{2.104}$$

$$\langle x, y \rangle_s := \lim_{t \to 0+} \frac{\|y + tx\|^2 - \|y\|^2}{2t}$$

where x, y are vectors in E. The mappings $\langle \cdot, \cdot \rangle_s$ and $\langle \cdot, \cdot \rangle_i$ are called the *superior* respectively the *inferior semi-inner product* associated with the norm $\|\cdot\|$.

For the sake of completeness we list here some properties of $\langle \cdot, \cdot \rangle_{s(i)}$ that will be used in the sequel.

We have, for $p, q \in \{i, s\}$ and $p \neq q$, that

(i) $\langle x, x \rangle_p = \|x\|^2$ for any $x \in E$;
(ii) $\langle ix, x \rangle_p = \langle x, ix \rangle_p = 0$ for any $x \in E$;
(iii) $\langle \lambda x, y \rangle_p = \lambda \langle x, y \rangle_p = \langle x, \lambda y \rangle_p$ for any $\lambda \geq 0$ and $x, y \in E$;
(iv) $\langle -x, y \rangle_p = \langle x, -y \rangle_p = -\langle x, y \rangle_q$ for any $x, y \in E$;
(v) $\langle ix, y \rangle_p = -\langle x, iy \rangle_p$ for any $x, y \in E$;
(vi) the following Schwarz type inequality holds:

$$\left| \langle x, y \rangle_p \right| \leq \|x\| \|y\|,$$

for any $x, y \in E$;

(vii) the following identity holds:

$$\langle \alpha x + y, x \rangle_p = \alpha \|x\|^2 + \langle y, x \rangle_p,$$

for any $\alpha \in \mathbb{R}$ and $x, y \in E$;

(viii) the following sub(super)-additivity property holds:

$$\langle x + y, z \rangle_p \leq (\geq) \langle x, z \rangle_p + \langle y, z \rangle_p,$$

for any $x, y, z \in E$, where the sign "$\leq$" applies for the superior semi-inner product, while the sign "$\geq$" applies for the inferior one;

(ix) the following continuity property is valid:

$$\left| \langle x + y, z \rangle_p - \langle y, z \rangle_p \right| \leq \|x\| \|z\|,$$

for any $x, y, z \in E$;

(x) from the definition we have the inequality

$$\langle x, y \rangle_i \leq \langle x, y \rangle_s$$

for any $x, y \in E$.

In the Banach algebra $B(H)$ we can associate to both the operator norm $\|\cdot\|$ and the numerical radius $w(\cdot)$ the following semi-inner products:

$$\langle A, B \rangle_{s(i),n} := \lim_{t \to 0+(-)} \frac{\|B + tA\|^2 - \|B\|^2}{2t} \tag{2.105}$$

and

$$\langle A, B \rangle_{s(i),w} := \lim_{t \to 0+(-)} \frac{w^2(B + tA) - w^2(B)}{2t} \tag{2.106}$$

respectively, where $A, B \in B(H)$.

It is obvious that the semi-inner products $\langle \cdot, \cdot \rangle_{s(i),n(w)}$ defined above have the usual properties of such mappings defined on general normed spaces and some special properties that will be specified in the following.

As a specific property that follows by the well-known inequality between the norm and the numerical radius of an operator, i.e., $w(T) \leq \|T\|$ for any $T \in B(H)$, we have

$$\langle T, I \rangle_{i,n} \leq \langle T, I \rangle_{i,w} (\leq) \langle T, I \rangle_{s,w} \leq \langle T, I \rangle_{s,n} \tag{2.107}$$

for any $T \in B(H)$, where I is the identity operator on H. We also observe that

$$\langle T, I \rangle_{s(i),n} = \lim_{t \to 0+(-)} \frac{\|I + tT\| - 1}{t}$$

and

$$\langle T, I \rangle_{s(i),w} = \lim_{t \to 0+(-)} \frac{w(I + tT) - 1}{t}$$

for any $T \in B(H)$.

It may be of interest to note that $\langle T, I \rangle_{s,n}$ and $\langle T, I \rangle_{s,w}$ are also called the *logarithmic norms* of T corresponding to $\|\cdot\|$ and w, respectively. Logarithmic norms corresponding to a given norm have been rather widely studied (mainly in the finite-dimensional case); see [53].

The following result is due to Lumer and was obtained originally for the numerical radius of operators in Banach spaces:

Theorem 65 (Lumer [42], 1961). *If $T \in B(H)$, then $\langle T, I \rangle_{p,n} = v_p(T)$, $p \in \{s, i\}$.*

The following simple result provides a connection between the semi-inner products generated by the operator norm and by the numerical radius as follows:

Theorem 66 (Dragomir [18], 2008). *For any $T \in B(H)$, we have*

$$\langle T, I \rangle_{p,n} = \langle T, I \rangle_{p,w}, \tag{2.108}$$

where $p \in \{s, i\}$.

Proof. Let us give a short proof for the case $p = s$.

Suppose $x \in H$, $\|x\| = 1$. Then for $t > 0$ we obviously have

$$\mathrm{Re}\,\langle Tx, x \rangle = \frac{\mathrm{Re}\,\langle x + tTx, x \rangle - 1}{t}$$
$$\leq \frac{|\langle x + tTx, x \rangle| - 1}{t} \leq \frac{w(I + tT) - 1}{t}. \tag{2.109}$$

Taking the supremum over $x \in H$, $\|x\| = 1$, we get

$$v_s(T) = \sup_{\|x\|=1} \mathrm{Re}\,\langle Tx, x \rangle \leq \frac{w(I + tT) - 1}{t}$$

for any $t > 0$, which implies, by letting $t \to 0+$ that

$$\sup_{\|x\|=1} \mathrm{Re}\,\langle Tx, x \rangle \leq \langle T, I \rangle_{s,w}, \tag{2.110}$$

for any $T \in B(H)$.

By Lumer's theorem we deduce then $\langle T, I \rangle_{s,n} \leq \langle T, I \rangle_{s,w}$ and since, by (2.107), we have $\langle T, I \rangle_{s,w} \leq \langle T, I \rangle_{s,n}$ the equality (2.108) is obtained. ∎

Now, on employing the properties of the semi-inner products outlined above, we can state the following properties as well:

(va) $v_{s(i)}(T) = \langle T, I \rangle_{s(i),w}$ for any $T \in B(H)$;
(vaa) $v_{s(i)}(T) = v_{s(i)}(\alpha I + T) - \alpha$ for any $\alpha \in \mathbb{R}$ and $T \in B(H)$;
(vaaa) $\left| v_{s(i)}(T + B) - v_{s(i)}(B) \right| \le w(T)$ for any $T, B \in B(H)$.

The following inequalities may be stated as well:

Theorem 67 (Dragomir [18], 2008). *For any $T \in B(H)$ and $\lambda \in \mathbb{C}$ we have*

$$\frac{1}{2}\left[\|T\|^2 + |\lambda|^2\right] \ge v_s\left(\bar{\lambda}T\right)$$

$$\ge \begin{cases} \frac{1}{2}\left[\|T\|^2 + |\lambda|^2\right] - \frac{1}{2}\|T - \lambda I\|^2, \\ \frac{1}{4}\left[\|T + \lambda I\|^2 - \|T - \lambda I\|^2\right], \end{cases} \tag{2.111}$$

and

$$\frac{1}{2}\left[w^2(T) + |\lambda|^2\right] \ge v_s\left(\bar{\lambda}T\right)$$

$$\ge \begin{cases} \frac{1}{2}\left[w^2(T) + |\lambda|^2\right] - \frac{1}{2}w^2(T - \lambda I), \\ \frac{1}{4}\left[w^2(T + \lambda I) - w^2(T - \lambda I)\right]. \end{cases} \tag{2.112}$$

respectively.

Proof. Let $x \in H$, $\|x\| = 1$. Then, obviously

$$0 \le \|Tx\|^2 - 2\operatorname{Re}\left[\langle \bar{\lambda}Tx, x \rangle\right] + |\lambda|^2 = \|(T - \lambda I)x\|^2 \le \|T - \lambda I\|^2,$$

which is equivalent with

$$\frac{1}{2}\left[\|Tx\|^2 + |\lambda|^2\right] - \frac{1}{2}\|T - \lambda I\|^2 \le \operatorname{Re}\langle \bar{\lambda}Tx, x \rangle$$

$$\le \frac{1}{2}\left[\|Tx\|^2 + |\lambda|^2\right], \tag{2.113}$$

for any $x \in H$, $\|x\| = 1$.

Taking the supremum over $\|x\| = 1$ we get the first inequality in (2.111) and the one from the first branch in the second.

For $x \in H$, $\|x\| = 1$ we also have that

$$\|Tx + \lambda x\|^2 = \|Tx - \lambda x\|^2 + 4\operatorname{Re}\langle \bar{\lambda}Tx, x \rangle, \tag{2.114}$$

which, on taking the supremum over $\|x\| = 1$, will produce the second part of the second inequality in (2.111).

The second inequality may be proven in a similar way. The details are omitted. ∎

It is well known, in general, that the semi-inner products $\langle \cdot, \cdot \rangle_{s(i)}$ defined on Banach spaces are not commutative. However, for the Banach algebra $B(H)$, we can point out the following connection between $\langle I, T \rangle_{s(i),n(w)}$ and the quantities $v_i(T)$ and $v_s(T)$, where $T \in B(H)$.

Corollary 68. *For any $T \in B(H)$ we have*

$$(v_i(T) =) \langle T, I \rangle_{i,n} \leq \frac{1}{2} \left[\langle I, T \rangle_{s,n} + \langle I, T \rangle_{i,n} \right]$$

$$\leq \langle T, I \rangle_{s,n} \, (= v_s(T)) . \tag{2.115}$$

and

$$(v_i(T) =) \langle T, I \rangle_{i,w} \leq \frac{1}{2} \left[\langle I, T \rangle_{s,w} + \langle I, T \rangle_{i,w} \right]$$

$$\leq \langle T, I \rangle_{s,w} \, (= v_s(T)) . \tag{2.116}$$

Proof. We have from the second part of the second inequality in (2.111) that

$$\frac{1}{2} \left[\frac{\|T + tI\|^2 - \|T\|^2}{2t} - \frac{\|T - tI\|^2 - \|T\|^2}{2t} \right] \leq v_s(T) \tag{2.117}$$

for any $t > 0$.

Taking the limit over $t \to 0+$ and noticing that

$$\lim_{t \to 0+} \frac{\|T - tI\|^2 - \|T\|^2}{2t} = \langle -I, T \rangle_{s,n} = -\langle I, T \rangle_{i,n} ,$$

we get the second inequality in (2.115).

Now, writing the second inequality in (2.115) for $-T$, we get

$$v_s(-T) \geq \frac{1}{2} \left[\langle I, -T \rangle_{s,n} + \langle I, -T \rangle_{i,n} \right]$$

$$= -\frac{1}{2} \left[\langle I, T \rangle_{s,n} + \langle I, T \rangle_{i,n} \right] ,$$

which is equivalent with the first part of (2.115).   ∎

Since $w(T) \leq \|T\|$, hence the first inequality in (2.112) provides a better upper bound for $v_s(\bar{\lambda} T)$ than the first inequality in (2.111).

2.4.3 Reverse Inequalities in Terms of the Operator Norm

The following result concerning reverse inequalities for the maximum of the spectrum of the real part and the operator norm of $T \in B(H)$ may be stated:

Theorem 69 (Dragomir [18], 2008). *For any $T \in B(H) \setminus \{0\}$ and $\lambda \in \mathbb{C} \setminus \{0\}$ we have the inequality:*

$$(0 \le \|T\| - w(T) \le) \|T\| - v_s \left(\frac{\bar{\lambda}}{|\lambda|} T \right)$$

$$\le \frac{1}{2|\lambda|} \|T - \lambda I\|^2 . \tag{2.118}$$

In addition, if $\|T - \lambda I\| \le |\lambda|$, then we have

$$\sqrt{1 - \left\| \frac{1}{\lambda} T - I \right\|^2} \le \frac{v_s \left(\frac{\bar{\lambda}}{|\lambda|} T \right)}{\|T\|} \left(\le \frac{w(T)}{\|T\|} \le 1 \right) \tag{2.119}$$

and

$$\left(0 \le \|T\|^2 - w^2(T) \le \right) \|T\|^2 - v_s^2 \left(\frac{\bar{\lambda}}{|\lambda|} T \right)$$

$$\le 2 \left(|\lambda| - \sqrt{|\lambda|^2 - \|T - \lambda I\|^2} \right) v_s \left(\frac{\bar{\lambda}}{|\lambda|} T \right)$$

$$\left(\le 2 \left(|\lambda| - \sqrt{|\lambda|^2 - \|T - \lambda I\|^2} \right) w(T) \right), \tag{2.120}$$

respectively.

Proof. Utilizing the property (av), we have

$$w(T) = w\left(\frac{\bar{\lambda}}{|\lambda|} T \right) \ge \left| v_s \left(\frac{\bar{\lambda}}{|\lambda|} T \right) \right| \ge v_s \left(\frac{\bar{\lambda}}{|\lambda|} T \right),$$

for any $\lambda \in \mathbb{C} \setminus \{0\}$ and the first inequality in (2.118) is proved.

By the arithmetic mean-geometric mean inequality we have

$$\frac{1}{2} \left[\|T\|^2 + |\lambda|^2 \right] \ge |\lambda| \|T\|,$$

which, by (2.111), provides

$$v_s \left(\bar{\lambda} T \right) \ge |\lambda| \|T\| - \frac{1}{2} \|T - \lambda I\|^2$$

that is equivalent with the second inequality in (2.118).

Utilizing the second part of the inequality (2.111) and under the assumption that $\|T - \lambda I\| \le |\lambda|$ we can also state that

$$v_s\left(\bar{\lambda}T\right) \ge \frac{1}{2}\left[\|T\|^2 + \left(\sqrt{|\lambda|^2 - \|T - \lambda I\|^2}\right)^2\right]. \tag{2.121}$$

By the arithmetic mean-geometric mean inequality we have now

$$\frac{1}{2}\left[\|T\|^2 + \left(\sqrt{|\lambda|^2 - \|T - \lambda I\|^2}\right)^2\right]$$

$$\ge \|T\|\sqrt{|\lambda|^2 - \|T - \lambda I\|^2}, \tag{2.122}$$

which, together with (2.121), implies the first inequality in (2.119).

The second part of (2.119) follows from (av).

From the proof of Theorem 67 we can state that

$$\|Tx\|^2 + |\lambda|^2 \le 2\,\mathrm{Re}\left(\bar{\lambda}Tx, x\right) + r^2, \qquad \|x\| = 1 \tag{2.123}$$

where we denoted $r := \|T - \lambda I\| \le |\lambda|$. We also observe, from (2.123), that $\mathrm{Re}\left(\bar{\lambda}Tx, x\right) > 0$ for $x \in H$, $\|x\| = 1$.

Now, if we divide (2.123) by $\mathrm{Re}\left(\frac{\bar{\lambda}}{|\lambda|}Tx, x\right) > 0$, we get

$$\frac{\|Tx\|^2}{\mathrm{Re}\left(\frac{\bar{\lambda}}{|\lambda|}Tx, x\right)} \le 2|\lambda| + \frac{r^2}{\mathrm{Re}\left(\frac{\bar{\lambda}}{|\lambda|}Tx, x\right)} - \frac{|\lambda|^2}{\mathrm{Re}\left(\frac{\bar{\lambda}}{|\lambda|}Tx, x\right)} \tag{2.124}$$

for $\|x\| = 1$.

If in this inequality we subtract from both sides the quantity $\mathrm{Re}\left(\frac{\bar{\lambda}}{|\lambda|}Tx, x\right)$, then we get

$$\frac{\|Tx\|^2}{\mathrm{Re}\left(\frac{\bar{\lambda}}{|\lambda|}Tx, x\right)} - \mathrm{Re}\left(\frac{\bar{\lambda}}{|\lambda|}Tx, x\right)$$

$$\le 2|\lambda| + \frac{r^2 - |\lambda|^2}{\mathrm{Re}\left(\frac{\bar{\lambda}}{|\lambda|}Tx, x\right)} - \mathrm{Re}\left(\frac{\bar{\lambda}}{|\lambda|}Tx, x\right)$$

$$= 2|\lambda| - \left(\frac{\sqrt{|\lambda|^2 - r^2}}{\sqrt{\mathrm{Re}\left(\frac{\bar{\lambda}}{|\lambda|}Tx, x\right)}} - \sqrt{\mathrm{Re}\left(\frac{\bar{\lambda}}{|\lambda|}Tx, x\right)}\right)^2 - 2\sqrt{|\lambda|^2 - r^2}$$

$$\le 2\left(|\lambda| - \sqrt{|\lambda|^2 - r^2}\right),$$

which obviously implies that

$$\|Tx\|^2 \le \left(\mathrm{Re} \left\langle \frac{\bar{\lambda}}{|\lambda|} Tx, x \right\rangle \right)^2$$
$$+ 2 \left(|\lambda| - \sqrt{|\lambda|^2 - r^2} \right) \mathrm{Re} \left\langle \frac{\bar{\lambda}}{|\lambda|} Tx, x \right\rangle \qquad (2.125)$$

for any $x \in H$, $\|x\| = 1$.

Now, taking the supremum in (2.125) over $x \in H$, $\|x\| = 1$, we deduce the second inequality in (2.120). The other inequalities are obvious and the theorem is proved. ∎

The following lemma is of interest in itself.

Lemma 70 (Dragomir [18], 2008). *For any $T \in B(H)$ and $\gamma, \Gamma \in \mathbb{C}$ we have*

$$v_i \left[(T^* - \bar{\gamma} I)(\Gamma I - T) \right] = \frac{1}{4} |\Gamma - \gamma|^2 - \left\| T - \frac{\gamma + \Gamma}{2} \cdot I \right\|^2. \qquad (2.126)$$

Proof. We observe that, for any $u, v, y \in H$, we have

$$\mathrm{Re} \langle u - y, y - v \rangle = \frac{1}{4} \|u - v\|^2 - \left\| y - \frac{u + v}{2} \right\|^2. \qquad (2.127)$$

Now, choosing $u = \Gamma x$, $y = Tx$, $v = \gamma x$ with $x \in H$, $\|x\| = 1$, we get

$$\mathrm{Re} \langle \Gamma x - Tx, Tx - \gamma x \rangle$$
$$= \frac{1}{4} |\Gamma - \gamma|^2 - \left\| Tx - \frac{\gamma + \Gamma}{2} x \right\|^2,$$

giving

$$\inf_{\|x\|=1} \mathrm{Re} \left\langle (T^* - \bar{\gamma} I)(\Gamma I - T) x, x \right\rangle$$
$$= \frac{1}{4} |\Gamma - \gamma|^2 - \sup_{\|x\|=1} \left\| Tx - \frac{\gamma + \Gamma}{2} x \right\|^2,$$

which is equivalent with (2.126). ∎

The following result providing a characterization for a class of operators that will be used in the sequel is incorporated in:

Lemma 71. *For $T \in B(H)$, $\gamma, \Gamma \in \mathbb{C}$ with $\Gamma \ne \gamma$ and $q \in \mathbb{R}$, the following statements are equivalent:*

(i) the operator $(T^* - \bar{\gamma} I)(\Gamma I - T)$ is q^2-accretive;

(ii) we have the norm inequality:

$$\left\| T - \frac{\gamma + \Gamma}{2} \cdot I \right\|^2 \leq \frac{1}{4} |\Gamma - \gamma|^2 - q^2. \tag{2.128}$$

The proof is obvious by Lemma 70 and the details are omitted.

Since the self-adjoint operators B satisfying the condition $B \geq mI$ in the operator partial under "$\geq$"are m−accretive, then, a sufficient condition for $C_{\gamma,\Gamma}(T) := (T^* - \bar{\gamma}I)(\Gamma I - T)$ to be q^2−accretive is that $C_{\gamma,\Gamma}(T)$ is self-adjoint and $C_{\gamma,\Gamma}(T) \geq q^2 I$.

Corollary 72. *Let* $T \in B(H)$, $\gamma, \Gamma \in \mathbb{C}$ *with* $\Gamma \neq \pm\gamma$ *and* $q \in \mathbb{R}$. *If the operator* $C_{\gamma,\Gamma}(T)$ *is* q^2−*accretive, then*

$$(0 \leq \|T\| - w(T) \leq) \|T\| - v_s \left(\frac{\bar{\Gamma} + \bar{\gamma}}{|\Gamma + \gamma|} \cdot T \right)$$

$$\leq \frac{1}{|\gamma + \Gamma|} \left[\frac{1}{4} |\Gamma - \gamma|^2 - q^2 \right]. \tag{2.129}$$

If M, m are positive real numbers with $M > m$ and the operator $C_{m,M}(T) = (T^* - mI)(MI - T)$ is q^2−accretive, then

$$(0 \leq \|T\| - w(T) \leq) \|T\| - v_s(T)$$

$$\leq \frac{1}{M + m} \left[\frac{1}{4}(M - m)^2 - q^2 \right]. \tag{2.130}$$

We observe that for $q = 0$, i.e., if $C_{\gamma,\Gamma}(T)$ respectively $C_{m,M}(T)$ are accretive, then we obtain from (2.129) and (2.130) the inequalities:

$$(0 \leq \|T\| - w(T) \leq) \|T\| - v_s \left(\frac{\bar{\Gamma} + \bar{\gamma}}{|\Gamma + \gamma|} \cdot T \right) \leq \frac{|\Gamma - \gamma|^2}{4|\Gamma + \gamma|} \tag{2.131}$$

and

$$(0 \leq \|T\| - w(T) \leq) \|T\| - v_s(T) \leq \frac{(M - m)^2}{4(M + m)} \tag{2.132}$$

respectively, which provide refinements of the corresponding inequalities (2.7) and (2.34) from [13].

For any bounded linear operator T we know that $\frac{w(T)}{\|T\|} \geq \frac{1}{2}$; therefore (2.119) would produce a useful result only if

$$\frac{1}{2} \leq \sqrt{1 - \left\| \frac{1}{\lambda} T - I \right\|^2},$$

which is equivalent with

$$\|T - \lambda I\| \leq \frac{\sqrt{3}}{2} |\lambda|. \tag{2.133}$$

In conclusion, for $T \in B(H) \setminus \{0\}$ and $\lambda \in \mathbb{C} \setminus \{0\}$ satisfying the condition (2.133), the inequality (2.119) provides a refinement of the classical result:

$$\frac{1}{2} \leq \frac{w(T)}{\|T\|}, \qquad T \in B(H). \tag{2.134}$$

Corollary 73. *Assume that* $\lambda \neq 0$ *(or* $T \neq 0$*). If* $\|T - \lambda I\| \leq |\lambda|$*, then we have*

$$\left(0 \leq \|T\|^2 - w^2(T) \leq\right) \|T\|^2 - v_s^2\left(\frac{\bar{\lambda}}{|\lambda|} \cdot T\right)$$

$$\leq \frac{\|T\|^2 \|T - \lambda I\|^2}{|\lambda|^2}. \tag{2.135}$$

The proof follows by the inequality (2.119). The details are omitted.
The following corollary may be stated as well:

Corollary 74. *Let* $T \in B(H) \setminus \{0\}$ *and* $\gamma, \Gamma \in \mathbb{C}$, $\Gamma \neq -\gamma$, $q \in \mathbb{R}$ *so that* $\mathrm{Re}(\Gamma \bar{\gamma}) + q^2 \geq 0$. *If* $C_{\gamma,\Gamma}(T)$ *is* q^2*–accretive, then*

$$\frac{2\sqrt{\mathrm{Re}(\Gamma \bar{\gamma}) + q^2}}{|\Gamma + \gamma|} \leq \frac{v_s\left(\frac{\bar{\Gamma}+\bar{\gamma}}{|\Gamma+\gamma|} \cdot T\right)}{\|T\|} \left(\leq \frac{w(T)}{\|T\|} \leq 1\right) \tag{2.136}$$

and

$$\left(0 \leq \|T\|^2 - w^2(T) \leq\right) \|T\|^2 - v_s^2\left(\frac{\bar{\Gamma}+\bar{\gamma}}{|\Gamma+\gamma|} \cdot T\right)$$

$$\leq \frac{2\|T\|^2}{|\Gamma + \gamma|}\left[\frac{1}{4}|\Gamma - \gamma|^2 - q^2\right]\left(\leq \frac{\|T\|^2 |\Gamma - \gamma|^2}{2|\Gamma + \gamma|}\right). \tag{2.137}$$

If γ, Γ and q are such that $|\Gamma + \gamma| \leq 4\sqrt{\mathrm{Re}(\Gamma \bar{\gamma}) + q^2}$, then (2.136) will provide a refinement of the classical result (2.134).
If $M > m \geq 0$ and the operator $C_{m,M}(T)$ is q^2–accretive, then

$$\frac{2\sqrt{Mm + q^2}}{m + M} \leq \frac{v_s(T)}{\|T\|} \left(\leq \frac{w(T)}{\|T\|} \leq 1\right) \tag{2.138}$$

and

$$\left(0 \le \|T\|^2 - w^2\left(T\right) \le\right) \|T\|^2 - v_s^2\left(T\right)$$

$$\le \frac{2\|T\|^2}{m+M}\left[\frac{1}{4}\left(M-m\right)^2 - q^2\right]. \tag{2.139}$$

We also observe that, for $q = 0$, i.e., if $C_{\gamma,\Gamma}\left(T\right)$ respectively $C_{m,M}\left(T\right)$ are accretive, then we obtain:

$$\frac{2\sqrt{\mathrm{Re}\left(\Gamma\bar{\gamma}\right)}}{|\Gamma+\gamma|} \le \frac{v_s\left(\frac{\bar{\Gamma}+\bar{\gamma}}{|\Gamma+\gamma|}\cdot T\right)}{\|T\|}\left(\le \frac{w\left(T\right)}{\|T\|}\right), \tag{2.140}$$

$$\frac{2\sqrt{Mm}}{m+M} \le \frac{v_s\left(T\right)}{\|T\|}\left(\le \frac{w\left(T\right)}{\|T\|}\right), \tag{2.141}$$

$$\left(0 \le \|T\|^2 - w^2\left(T\right) \le\right) \|T\|^2 - v_s^2\left(\frac{\bar{\Gamma}+\bar{\gamma}}{|\Gamma+\gamma|}\cdot T\right)$$

$$\le \frac{\|T\|^2\,|\Gamma-\gamma|^2}{2\,|\Gamma+\gamma|} \tag{2.142}$$

and

$$\left(0 \le \|T\|^2 - w^2\left(T\right) \le\right) \|T\|^2 - v_s^2\left(T\right) \le \frac{\|T\|^2\left(M-m\right)^2}{2\left(m+M\right)} \tag{2.143}$$

respectively, which provides refinements of the inequalities (2.17), (2.31) and (2.20) from [13], respectively. The inequality between the first and the last term in (2.143) was not stated in [13].

Corollary 75. *Let $T \in B\left(H\right)$, $\gamma, \Gamma \in \mathbb{C}$, $\Gamma \ne -\gamma$, $q \in \mathbb{R}$ so that $\mathrm{Re}\left(\Gamma\bar{\gamma}\right) + q^2 \ge 0$. If $C_{\gamma,\Gamma}\left(H\right)$ is q^2-accretive, then*

$$\left(0 \le \|T\|^2 - w^2\left(T\right) \le\right) \|T\|^2 - v_s^2\left(\frac{\bar{\Gamma}+\bar{\gamma}}{|\Gamma+\gamma|}\cdot T\right)$$

$$\le \left(|\Gamma+\gamma| - 2\sqrt{\mathrm{Re}\left(\Gamma\bar{\gamma}\right)+q^2}\right) v_s\left(\frac{\bar{\Gamma}+\bar{\gamma}}{|\Gamma+\gamma|}\cdot T\right)$$

$$\left(\le \left(|\Gamma+\gamma| - 2\sqrt{\mathrm{Re}\left(\Gamma\bar{\gamma}\right)+q^2}\right) w\left(T\right)\right). \tag{2.144}$$

The proof follows by the last part of Theorem 69. The details are omitted.

If $M > m \geq 0$ and the operator $C_{m,M}(T)$ is q^2–accretive, then

$$(0 \leq \|T\|^2 - w^2(T) \leq) \|T\|^2 - v_s^2(T)$$

$$\leq \left(M + m - 2\sqrt{Mm + q^2}\right) v_s(T)$$

$$\left(\leq \left(M + m - 2\sqrt{Mm + q^2}\right) w(T)\right). \qquad (2.145)$$

Finally, for $q = 0$, i.e., if $C_{\gamma,\Gamma}(T)$ respectively $C_{m,M}(T)$ are accretive, then we obtain from (2.144) and (2.145) some refinements of the inequalities (2.29) and (2.33) from [13].

2.4.4 Reverse Inequalities in Terms of the Numerical Radius

It is well known that the following lower bound for the numerical radius $w(T)$ holds (see (av) from Introduction)

$$\left|v_p(T)\right| \leq w(T), \ p \in \{s, i\}, \qquad (2.146)$$

for any T a bounded linear operator, where, as in the introduction,

$$v_{s(i)}(T) = \langle T, I \rangle_{s(i)} = \sup_{\|x\|=1} \left(\inf_{\|x\|=1}\right) \mathrm{Re}\langle Tx, x \rangle \qquad (2.147)$$

$$= \lambda_{\max(\min)}(\mathrm{Re}(T)).$$

It is then a natural problem to investigate how far the left side of (2.146) from the numerical radius is?

We start with the following result:

Theorem 76 (Dragomir [18], 2008). *For any $T \in B(H) \backslash \{0\}$ and $\lambda \in \mathbb{C} \backslash \{0\}$ we have*

$$\left(0 \leq w(T) - \left|v_s\left(\frac{\bar{\lambda}}{|\lambda|}T\right)\right| \leq\right) w(T) - v_s\left(\frac{\bar{\lambda}}{|\lambda|}T\right)$$

$$\leq \frac{1}{2|\lambda|} w^2(T - \lambda I) \left(\leq \frac{1}{2|\lambda|} \|T - \lambda I\|^2\right). \qquad (2.148)$$

Moreover, if $w(T - \lambda I) \leq |\lambda|$, then we have

$$\sqrt{1 - w^2\left(\frac{1}{\lambda}T - I\right)} \leq \frac{v_s\left(\frac{\bar{\lambda}}{|\lambda|}T\right)}{w(T)} \left(\leq \frac{\left|v_s\left(\frac{\bar{\lambda}}{|\lambda|}T\right)\right|}{w(T)} \leq 1\right) \qquad (2.149)$$

and

$$(0 \leq) w^2 (T) - v_s^2 \left(\frac{\bar{\lambda}}{|\lambda|} T \right)$$

$$\leq 2 \left(|\lambda| - \sqrt{|\lambda|^2 - w^2 (T - \lambda I)} \right) v_s \left(\frac{\bar{\lambda}}{|\lambda|} T \right)$$

$$\left(\leq 2 \left(|\lambda| - \sqrt{|\lambda|^2 - w^2 (T - \lambda I)} \right) w (T) \right), \tag{2.150}$$

respectively.

Proof. The argument is similar with the one from Theorem 69 and the details are omitted.   ∎

The following lemma is of interest.

Lemma 77 (Dragomir [18], 2008). *For any $T \in B(H)$ and $\gamma, \Gamma \in \mathbb{C}$ we have*

$$\inf_{\|x\|=1} \mathrm{Re} \left[\langle (\Gamma I - T) x, x \rangle \langle x, (T - \gamma I) x \rangle \right]$$

$$= \frac{1}{4} |\Gamma - \gamma|^2 - w^2 \left(T - \frac{\gamma + \Gamma}{2} \cdot I \right). \tag{2.151}$$

Proof. We observe that for any u, v, y complex numbers, we have the elementary identity:

$$\mathrm{Re} \left[(u - y)(\bar{y} - \bar{v}) \right] = \frac{1}{4} |u - v|^2 - \left| y - \frac{u + v}{2} \right|^2. \tag{2.152}$$

If we choose in (2.152) $u = \Gamma$, $y = \langle Tx, x \rangle$ and $v = \gamma$ with $x \in H$, $\|x\| = 1$, then by (2.152) we have

$$\mathrm{Re} \left[\langle (\Gamma I - T) x, x \rangle \langle x, (T - \gamma I) x \rangle \right]$$

$$= \frac{1}{4} |\Gamma - \gamma|^2 - \left| \left\langle \left(T - \frac{\gamma + \Gamma}{2} \cdot I \right) x, x \right\rangle \right|^2 \tag{2.153}$$

for any $x \in H$, $\|x\| = 1$.

Now, taking the infimum over $\|x\| = 1$ in (2.153), we deduce the desired identity (2.151).   ∎

We observe that for any $x \in H$, $\|x\| = 1$ we have

$$\mu (T; \gamma, \Gamma) (x) := \mathrm{Re} \left[\langle (\Gamma I - T) x, x \rangle \langle x, (T - \gamma I) x \rangle \right]$$

$$= (\mathrm{Re}\, \Gamma - \mathrm{Re}\, \langle Tx, x \rangle) (\mathrm{Re}\, \langle Tx, x \rangle - \mathrm{Re}\, \gamma)$$

$$+ (\mathrm{Im}\, \Gamma - \mathrm{Im}\, \langle Tx, x \rangle) (\mathrm{Im}\, \langle Tx, x \rangle - \mathrm{Im}\, \gamma)$$

and therefore a sufficient condition for $\mu\left(T;\gamma,\Gamma\right)(x)$ to be nonnegative for any $x \in H, \|x\| = 1$ is that:

$$\begin{cases} \operatorname{Re}\Gamma \geq \operatorname{Re}\langle Tx,x\rangle \geq \operatorname{Re}\gamma, \\ \operatorname{Im}\Gamma \geq \operatorname{Im}\langle Tx,x\rangle \geq \operatorname{Im}\gamma, \end{cases} \quad x \in H, \|x\| = 1. \quad (2.154)$$

Now, if we denote by $\mu_i\left(T;\gamma,\Gamma\right) := \inf_{\|x\|=1}\mu\left(T;\gamma,\Gamma\right)(x)$, then we can state the following lemma.

Lemma 78 (Dragomir [18], 2008). *For $T \in B(H)$, $\phi, \Phi \in \mathbb{C}$, the following statements are equivalent:*

(i) $\mu_i\left(T;\phi;\Phi\right) \geq 0$;
(ii) $w\left(T - \frac{\phi+\Phi}{2}\cdot I\right) \leq \frac{1}{2}|\Phi - \phi|$.

Utilizing the above results we can provide now some particular reverse inequalities that are of interest.

Corollary 79. *Let $T \in B(H)$ and $\phi, \Phi \in \mathbb{C}$ with $\Phi \neq \pm\phi$ such that either (i) or (ii) of Lemma 78 holds true. Then*

$$\left(0 \leq w(T) - \left|v_s\left(\frac{\bar{\phi}+\bar{\Phi}}{|\phi+\Phi|}\cdot T\right)\right|\leq\right) w(T) - v_s\left(\frac{\bar{\phi}+\bar{\Phi}}{|\phi+\Phi|}\cdot T\right)$$

$$\leq \frac{1}{4}\cdot\frac{|\Phi-\phi|^2}{|\Phi+\phi|}. \quad (2.155)$$

If $N > n > 0$ are such that either $\mu_i\left(T;n,N\right) \geq 0$ or $w\left(T - \frac{n+N}{2}\cdot I\right) \leq \frac{1}{2}\left(N - n\right)$ for a given operator $T \in B(T)$, then

$$(0 \leq w(T) - |v_s(T)| \leq)w(T) - v_s(T) \leq \frac{1}{4}\cdot\frac{(N-n)^2}{N+n}. \quad (2.156)$$

An equivalent additive version of (2.149) is incorporated in the following:

Corollary 80. *Assume that $\lambda \neq 0$ (or $T \neq 0$). If $w\left(T - \lambda I\right) \leq |\lambda|$, then we have*

$$(0 \leq)w^2(T) - w_s^2\left(\frac{\bar{\lambda}}{|\lambda|}T\right) \leq \frac{w^2(T)w^2(T-\lambda I)}{|\lambda|^2}$$

$$\left(\leq \begin{cases} \frac{\|T\|^2 w^2(T-\lambda I)}{|\lambda|^2} \\ \frac{w^2(T)\|T-\lambda I\|^2}{|\lambda|^2} \end{cases} \leq \frac{\|T\|^2\|T-\lambda I\|^2}{|\lambda|^2}\right). \quad (2.157)$$

The following variant of (2.149) can be perhaps more convenient:

Corollary 81. *Let $T \in B(H) \setminus \{0\}$ and $\phi, \Phi \in \mathbb{C}$ with $\Phi \neq -\phi$. If $\mathrm{Re}\left(\Phi\bar{\phi}\right) > 0$ and either the statement (i) or equivalently (ii) from Lemma 78 holds true, then:*

$$\frac{2\sqrt{\mathrm{Re}\left(\Phi\bar{\phi}\right)}}{|\phi + \Phi|} \leq \frac{v_s\left(\frac{\bar{\phi}+\bar{\Phi}}{|\phi+\Phi|} \cdot T\right)}{w(T)} \left(\leq \frac{\left|v_s\left(\frac{\bar{\phi}+\bar{\Phi}}{|\phi+\Phi|} \cdot T\right)\right|}{w(T)} \leq 1\right) \tag{2.158}$$

and

$$(0 \leq) w^2(T) - v_s^2\left(\frac{\bar{\phi} + \bar{\Phi}}{|\phi + \Phi|} \cdot T\right)$$

$$\leq \frac{1}{4} \cdot \frac{|\Phi - \phi|^2}{\mathrm{Re}\left(\Phi\bar{\phi}\right)} v_s^2\left(\frac{\bar{\phi} + \bar{\Phi}}{|\phi + \Phi|} \cdot T\right)$$

$$\left(\leq \frac{1}{4} \cdot \frac{|\Phi - \phi|^2}{\mathrm{Re}\left(\Phi\bar{\phi}\right)} w^2(T) \leq \frac{1}{4} \cdot \frac{|\Phi - \phi|^2}{\mathrm{Re}\left(\Phi\bar{\phi}\right)} \|T\|^2\right). \tag{2.159}$$

The proof follows by Theorem 76 and the details are omitted.

If $N > n > 0$ are such that either $\mu_i(T; n, N) \geq 0$ or, equivalently

$$w\left(T - \frac{n+N}{2} \cdot I\right) \leq \frac{1}{2}(N - n), \tag{2.160}$$

then

$$\frac{2\sqrt{nN}}{n+N} \leq \frac{v_s(T)}{w(T)} \ (\leq 1), \tag{2.161}$$

$$(0 \leq) w(T) - v_s(T) \leq \frac{\left(\sqrt{N} - \sqrt{n}\right)^2}{2\sqrt{nN}} v_s(T)$$

$$\left(\leq \frac{\left(\sqrt{N} - \sqrt{n}\right)^2}{2\sqrt{nN}} w(T)\right) \tag{2.162}$$

and

$$(0 \leq) w^2(T) - v_s^2(T) \leq \frac{(N - n)^2}{4nN} v_s^2(T) \tag{2.163}$$

$$\left(\leq \frac{(N - n)^2}{4nN} w^2(T)\right).$$

Finally, we can state the following result as well:

Corollary 82. *Let* $T \in B(H)$, $\phi, \Phi \in \mathbb{C}$ *such that* $\text{Re}(\Phi\bar{\phi}) > 0$. *If either* $\mu_i(T; \phi, \Phi) \geq 0$ *or, equivalently*

$$w\left(T - \frac{\Phi + \phi}{2} \cdot I\right) \leq \frac{1}{2}|\Phi - \phi|,$$

then

$$(0 \leq) w^2(T) - v_s^2\left(\frac{\bar{\phi} + \bar{\Phi}}{|\phi + \Phi|} \cdot T\right)$$

$$\leq \left[|\phi + \Phi| - 2\sqrt{\text{Re}(\Phi\bar{\phi})}\right] v_s\left(\frac{\bar{\phi} + \bar{\Phi}}{|\phi + \Phi|} \cdot T\right)$$

$$\left(\leq \left[|\phi + \Phi| - 2\sqrt{\text{Re}(\Phi\bar{\phi})}\right] w(T)\right). \tag{2.164}$$

Moreover, if $N > n > 0$ are such that $\mu_i(T; n, N) \geq 0$, then we have the simpler inequality:

$$(0 \leq) w^2(T) - v_s^2(T) \leq \left(\sqrt{N} - \sqrt{n}\right)^2 v_s(T)$$

$$\left(\leq \left(\sqrt{N} - \sqrt{n}\right)^2 w(T)\right). \tag{2.165}$$

2.5   New Inequalities of the Kantorovich Type

2.5.1   Some Classical Facts

Let $(H, \langle \cdot, \cdot \rangle)$ be a Hilbert space over the real or complex number field $\mathbb{K}$, $B(H)$ the $\mathbb{C}^*$-algebra of all bounded linear operators defined on H and $T \in B(H)$. If T is invertible, then we can define the *Kantorovich functional* as

$$K(T; x) := \langle Tx, x \rangle \langle T^{-1}x, x \rangle \tag{2.166}$$

for any $x \in H$, $\|x\| = 1$.

As pointed out by Greub and Rheinboldt in their seminal paper [32], if $M > m > 0$ and for the self-adjoint operator T, we have

$$MI \geq T \geq mI \tag{2.167}$$

in the partial operator order of $B(H)$, where I is the identity operator, then the *Kantorovich operator inequality* holds true

$$1 \le K\,(T;x) \le \frac{(M+m)^2}{4mM},\qquad (2.168)$$

for any $x \in H, \|x\| = 1$.

An equivalent *additive* form of this result is incorporated in:

$$0 \le K\,(T;x) - 1 \le \frac{(M-m)^2}{4mM},\qquad (2.169)$$

for any $x \in H, \|x\| = 1$.

For results related to the Kantorovich operator inequality we recommend the classical works of Strang [57], Diaz and Metcalf [3], Householder [35], Mond [44] and Mond and Shisha [47]. Other results have been obtained by Mond and Pečarić [45, 46], Fujii et al. [21, 22], Spain [54], Nakamoto and Nakamura [48], Furuta [25, 26], Tsukada and Takahasi [58] and more recently by Yamazaki [61], Furuta & Giga [27], Fujii and Nakamura [23, 24] and others.

Due to the important applications of the original Kantorovich inequality for matrices [36] in statistics [37, 40, 41, 50, 51, 55, 56, 59, 60, 62] and Numerical Analysis [1, 28–30, 52], any new inequality of this type will have a flow of consequences into other areas.

Motivated by interests in both pure and applied mathematics outlined above, we establish in this section some new inequalities of Kantorovich type. They are shown to hold for larger classes of operators and subsets of complex numbers than considered before in the literature and provide refinements of the classical result in the case when the involved operator T satisfies the usual condition (2.167). As natural tools in deriving the new results, the recent Grüss type inequalities for vectors in inner products obtained by the author in [5–7] are utilized. In the process, several new reverse inequalities for the numerical radius of a bounded linear operator are derived as well.

2.5.2 Some Grüss Type Inequalities

The following lemmas, that are of interest in their own right, collect some Grüss type inequalities for vectors in inner product spaces obtained earlier by the author:

Lemma 83. *Let* $(H, \langle \cdot, \cdot \rangle)$ *be an inner product space over the real or complex number field* $\mathbb{K}$, $u, v, e \in H$, $\|e\| = 1$, *and* $\alpha, \beta, \gamma, \delta \in \mathbb{K}$ *such that*

$$\mathrm{Re}\,\langle \beta e - u, u - \alpha e \rangle \ge 0, \mathrm{Re}\,\langle \delta e - v, v - \gamma e \rangle \ge 0 \qquad (2.170)$$

or equivalently,

$$\left\| u - \frac{\alpha + \beta}{2} e \right\| \leq \frac{1}{2} |\beta - \alpha|, \left\| v - \frac{\gamma + \delta}{2} e \right\| \leq \frac{1}{2} |\delta - \gamma|. \tag{2.171}$$

Then

$$|\langle u, v \rangle - \langle u, e \rangle \langle e, v \rangle| \tag{2.172}$$

$$\leq \frac{1}{4} |\beta - \alpha| |\delta - \gamma|$$

$$- \begin{cases} [\text{Re} \langle \beta e - u, u - \alpha e \rangle \, \text{Re} \langle \delta e - v, v - \gamma e \rangle]^{\frac{1}{2}}, \\ \left| \langle u, e \rangle - \frac{\alpha + \beta}{2} \right| \left| \langle v, e \rangle - \frac{\gamma + \delta}{2} \right|. \end{cases}$$

The first inequality has been obtained in [6] (see also [11, p. 44]) while the second result was established in [8] (see also [11, p. 90]). They provide refinements of the earlier result from [5] where only the first part of the bound, i.e., $\frac{1}{4} |\beta - \alpha| |\delta - \gamma|$ has been given. Notice that, as pointed out in [8], the upper bounds for the Grüss functional incorporated in (2.172) cannot be compared in general, meaning that one is better than the other depending on appropriate choices of the vectors and scalars involved.

Another result of this type is the following one:

Lemma 84. *With the assumptions in Lemma 83 and if* $\text{Re} \, (\beta \overline{\alpha}) > 0, \text{Re} \, (\delta \overline{\gamma}) > 0,$ *then*

$$|\langle u, v \rangle - \langle u, e \rangle \langle e, v \rangle| \tag{2.173}$$

$$\leq \begin{cases} \frac{1}{4} \frac{|\beta - \alpha| |\delta - \gamma|}{[\text{Re}(\beta \overline{\alpha}) \, \text{Re}(\delta \overline{\gamma})]^{\frac{1}{2}}} |\langle u, e \rangle \langle e, v \rangle|, \\ \left[\left(|\alpha + \beta| - 2 \, [\text{Re} \, (\beta \overline{\alpha})]^{\frac{1}{2}} \right) \left(|\delta + \gamma| - 2 \, [\text{Re} \, (\delta \overline{\gamma})]^{\frac{1}{2}} \right) \right]^{\frac{1}{2}} \\ \times [|\langle u, e \rangle \langle e, v \rangle|]^{\frac{1}{2}}. \end{cases}$$

The first inequality has been established in [9] (see [11, p. 62]) while the second one can be obtained in a canonical manner from the reverse of the Schwarz inequality given in [15]. The details are omitted.

Finally, another inequality of Grüss type that has been obtained in [7] (see also [11, p. 65]) can be stated as:

Lemma 85. *With the assumptions in Lemma 83 and if* $\beta \neq -\alpha, \delta \neq -\gamma,$ *then*

$$|\langle u, v \rangle - \langle u, e \rangle \langle e, v \rangle| \tag{2.174}$$

$$\leq \frac{1}{4} \frac{|\beta - \alpha| |\delta - \gamma|}{[|\beta + \alpha| |\delta + \gamma|]^{\frac{1}{2}}} [(\|u\| + |\langle u, e \rangle|)(\|v\| + |\langle v, e \rangle|)]^{\frac{1}{2}}.$$

2.5.3 Operator Inequalities of Grüss Type

For the complex numbers α, β and the bounded linear operator T we define the following transform

$$C_{\alpha,\beta}(T) := \left(T^* - \overline{\alpha}I\right)(\beta I - T), \tag{2.175}$$

where by T^* we denote the adjoint of T.

We list some properties of the transform $C_{\alpha,\beta}(\cdot)$ that are useful in the following:

(i) For any $\alpha, \beta \in \mathbb{C}$ and $T \in B(H)$ we have

$$C_{\alpha,\beta}(I) = (1 - \overline{\alpha})(\beta - 1)I, \tag{2.176}$$

$$C_{\alpha,\alpha}(T) = -(\alpha I - T)^*(\alpha I - T),$$

$$C_{\alpha,\beta}(\gamma T) = |\gamma|^2 C_{\frac{\alpha}{\gamma},\frac{\beta}{\gamma}}(T) \quad \text{for each } \gamma \in \mathbb{C}\backslash\{0\}, \tag{2.177}$$

$$\left[C_{\alpha,\beta}(T)\right]^* = C_{\beta,\alpha}(T) \tag{2.178}$$

and

$$C_{\overline{\beta},\overline{\alpha}}(T^*) - C_{\alpha,\beta}(T) = T^*T - TT^*. \tag{2.179}$$

(ii) The operator $T \in B(H)$ is normal if and only if $C_{\overline{\beta},\overline{\alpha}}(T^*) = C_{\alpha,\beta}(T)$ for each $\alpha, \beta \in \mathbb{C}$.

(iii) If $T \in B(H)$ is invertible and $\alpha, \beta \in \mathbb{C}\backslash\{0\}$, then

$$\left(T^{-1}\right)^* C_{\alpha,\beta}(T) T^{-1} = \overline{\alpha}\beta C_{\frac{1}{\alpha},\frac{1}{\beta}}\left(T^{-1}\right). \tag{2.180}$$

We recall that a bounded linear operator T on the complex Hilbert space $(H, \langle \cdot, \cdot \rangle)$ is called *accretive* if $\operatorname{Re}\langle Ty, y \rangle \geq 0$ for any $y \in H$.

The following simple characterization result is useful in the following:

Lemma 86. *For $\alpha, \beta \in \mathbb{C}$ and $T \in B(H)$ the following statements are equivalent:*

(i) *the transform $C_{\alpha,\beta}(T)$ is accretive;*
(ii) *the transform $C_{\overline{\alpha},\overline{\beta}}(T^*)$ is accretive;*
(iii) *we have the norm inequality*

$$\left\|T - \frac{\alpha + \beta}{2}I\right\| \leq \frac{1}{2}|\beta - \alpha|. \tag{2.181}$$

Proof. The proof of the equivalence "$(i) \Leftrightarrow (iii)$" is obvious by the equality

$$\operatorname{Re}\left\langle\left(T^* - \overline{\alpha}I\right)\left(\beta I - T\right)x, x\right\rangle = \frac{1}{4}\left|\beta - \alpha\right|^2 - \left\|\left(T - \frac{\alpha + \beta}{2}I\right)x\right\|^2 \quad (2.182)$$

which holds for any $\alpha, \beta \in \mathbb{C}$, $T \in B(H)$ and $x \in H$, $\|x\| = 1$. ∎

Remark 87. In order to give examples of operators $T \in B(H)$ and numbers $\alpha, \beta \in \mathbb{C}$ such that the transform $C_{\alpha,\beta}(T)$ is accretive, it suffices to select a bounded linear operator T and the complex numbers z, w with the property that $\|T - zI\| \leq |w|$ and, by choosing $T = T$, $\alpha = \frac{1}{2}(z + w)$ and $\beta = \frac{1}{2}(z - w)$ we observe that T satisfies (2.181), *i.e.* $C_{\alpha,\beta}(T)$ is accretive.

For two bounded linear operators $T, B \in B(H)$ and the vector $x \in H$, $\|x\| = 1$ define the functional

$$G(T, B; x) := \langle Tx, Bx \rangle - \langle Tx, x \rangle \langle x, Bx \rangle.$$

The following result concerning operator inequalities of Grüss type may be stated:

Theorem 88 (Dragomir [19], 2008). *Let $T, B \in B(H)$ and $\alpha, \beta, \gamma, \delta \in \mathbb{K}$ be such that the transforms $C_{\alpha,\beta}(T), C_{\gamma,\delta}(B)$ are accretive, then*

$$|G(T, B; x)| \quad (2.183)$$

$$\leq \frac{1}{4}|\beta - \alpha||\delta - \gamma|$$

$$- \begin{cases} \left[\operatorname{Re}\left\langle C_{\alpha,\beta}(T)x, x\right\rangle \operatorname{Re}\left\langle C_{\gamma,\delta}(B)x, x\right\rangle\right]^{\frac{1}{2}} \\ \left|\left\langle\left(T - \frac{\alpha+\beta}{2}I\right)x, x\right\rangle\right|\left|\left\langle\left(B - \frac{\gamma+\delta}{2}I\right)x, x\right\rangle\right| \end{cases}$$

$$\left(\leq \frac{1}{4}|\beta - \alpha||\delta - \gamma|\right),$$

for any $x \in H$, $\|x\| = 1$.
 If $\operatorname{Re}(\beta\overline{\alpha}) > 0$, $\operatorname{Re}(\delta\overline{\gamma}) > 0$, then

$$|G(T, B; x)| \quad (2.184)$$

$$\leq \begin{cases} \frac{1}{4}\frac{|\beta-\alpha||\delta-\gamma|}{[\operatorname{Re}(\beta\overline{\alpha})\operatorname{Re}(\delta\overline{\gamma})]^{\frac{1}{2}}}|\langle Tx, x \rangle \langle Bx, x \rangle|, \\ \left[\left(|\alpha + \beta| - 2\left[\operatorname{Re}(\beta\overline{\alpha})\right]^{\frac{1}{2}}\right)\left(|\delta + \gamma| - 2\left[\operatorname{Re}(\delta\overline{\gamma})\right]^{\frac{1}{2}}\right)\right]^{\frac{1}{2}} \\ \times \left[|\langle Tx, x \rangle \langle Bx, x \rangle|\right]^{\frac{1}{2}}, \end{cases}$$

for any $x \in H$, $\|x\| = 1$.

If $\beta \neq -\alpha, \delta \neq -\gamma$, then

$$|G(T, B; x)| \tag{2.185}$$

$$\leq \frac{1}{4} \frac{|\beta - \alpha| |\delta - \gamma|}{[|\beta + \alpha| |\delta + \gamma|]^{\frac{1}{2}}} [(\|Tx\| + |\langle Tx, x \rangle|) (\|Bx\| + |\langle Bx, x \rangle|)]^{\frac{1}{2}},$$

for any $x \in H, \|x\| = 1$.

The proof follows by Lemmas 83, 84 and 85 on choosing $u = Tx, v = Bx$ and $e = x, x \in H, \|x\| = 1$.

Remark 89. In order to give examples of operators $T \in B(H)$ and complex numbers α, β for which $C_{\alpha,\beta}(T)$ is accretive and $\text{Re}(\beta\bar{\alpha}) > 0$ it is enough to select in Remark 87 $z, w \in \mathbb{C}$ with $|z| > |w| > 0$. This follows from the fact that for $\alpha = \frac{1}{2}(z + w)$ and $\beta = \frac{1}{2}(z - w)$ we have $\text{Re}(\beta\bar{\alpha}) = \frac{1}{4}(|z|^2 - |w|^2)$.

Remark 90. We observe that

$$G(T, B^*; x) = \langle BTx, x \rangle - \langle Tx, x \rangle \langle Bx, x \rangle, x \in H, \|x\| = 1$$

and since, by Lemma 86, the transform $C_{\alpha,\beta}(T)$ is accretive if and only if $C_{\bar{\alpha},\bar{\beta}}(T^*)$ is accretive, hence in the inequalities (2.183)–(2.185) we can replace $G(T, B; x)$ by $G(T, B^*; x)$ to obtain other Grüss type inequalities that will be used in the sequel.

In some applications, the case $B = T$ in both quantities $G(T, B; x)$ and $G(T, B^*; x)$ may be of interest. For the sake of simplicity, we denote

$$G_1(T; x) := G(T, T; x) = \|Tx\|^2 - |\langle Tx, x \rangle|^2 \geq 0$$

and

$$G_2(T; x) := G(T, T^*; x) = \langle T^2x, x \rangle - [\langle Tx, x \rangle]^2,$$

for $x \in H, \|x\| = 1$. For these quantities, related to the Schwarz's inequality, we can state the following result which is of interest:

Corollary 91. *Let $T \in B(H)$ and $\alpha, \beta \in \mathbb{K}$ be such that the transform $C_{\alpha,\beta}(T)$ is accretive, then*

$$G_1(T; x) \tag{2.186}$$

$$\leq \frac{1}{4} |\beta - \alpha|^2 - \begin{cases} \text{Re} \langle C_{\alpha,\beta}(T) x, x \rangle \\ \left| \left\langle \left(T - \frac{\alpha+\beta}{2} I \right) x, x \right\rangle \right|^2 \end{cases} \left(\leq \frac{1}{4} |\beta - \alpha|^2 \right),$$

for any $x \in H, \|x\| = 1$.

If $\mathrm{Re}\,(\beta\overline{\alpha}) > 0$, *then*

$$G_1\,(T;x) \le \begin{cases} \frac{1}{4}\frac{|\beta-\alpha|^2}{\mathrm{Re}(\beta\overline{\alpha})}\,|\langle Tx,x\rangle|^2, \\[2mm] \left(|\alpha+\beta|-2\left[\mathrm{Re}\,(\beta\overline{\alpha})\right]^{\frac{1}{2}}\right)|\langle Tx,x\rangle|, \end{cases} \tag{2.187}$$

for any $x \in H$, $\|x\| = 1$.
   If $\beta \ne -\alpha$, *then*

$$G_1\,(T;x) \le \frac{1}{4}\frac{|\beta-\alpha|^2}{|\beta+\alpha|}\,(\|Tx\| + |\langle Tx,x\rangle|), \tag{2.188}$$

for any $x \in H$, $\|x\| = 1$.

A similar result holds for $G_2\,(T;x)$. The details are omitted.

2.5.4 Reverse Inequalities for the Numerical Range

It is well known that $w\,(\cdot)$ is a norm on the Banach algebra $B\,(H)$. This norm is equivalent with the operator norm. In fact, the following more precise result holds [33, p. 9]:

Theorem 92 (Equivalent norm). *For any* $T \in B\,(H)$ *one has*

$$w\,(T) \le \|T\| \le 2w\,(T). \tag{2.189}$$

The following reverses of the first inequality in (2.189), i.e., upper bounds under appropriate conditions for the bounded linear operator T for the nonnegative difference $\|T\|^2 - w^2\,(T)$, can be obtained:

Theorem 93 (Dragomir [19], 2008). *Let* $T \in B(H)$ *and* $\alpha, \beta \in \mathbb{K}$ *be such that the transform* $C_{\alpha,\beta}\,(T)$ *is accretive, then*

$$(0 \le)\,\|T\|^2 - w^2\,(T) \le \frac{1}{4}|\beta-\alpha|^2 - \begin{cases} \vartheta_i\left(C_{\alpha,\beta}\,(T)\right) \\[2mm] w_i^2\left(T - \frac{\alpha+\beta}{2}I\right) \end{cases} \tag{2.190}$$

$$\left(\le \frac{1}{4}|\beta-\alpha|^2\right),$$

where, for a given operator B, *we have denoted* $\vartheta_i\,(B) := \inf_{\|x\|=1} \mathrm{Re}\,\langle Tx,x\rangle$ *and* $w_i\,(B) := \inf_{\|x\|=1} |\langle Tx,x\rangle|$.
   If $\mathrm{Re}\,(\beta\overline{\alpha}) > 0$, *then*

$$(0 \le)\,\|T\|^2 - w^2\,(T) \le \begin{cases} \frac{1}{4}\frac{|\beta-\alpha|^2}{\mathrm{Re}(\beta\overline{\alpha})}w^2\,(T), \\[2mm] \left(|\alpha+\beta|-2\left[\mathrm{Re}\,(\beta\overline{\alpha})\right]^{\frac{1}{2}}\right)w\,(T). \end{cases} \tag{2.191}$$

If $\beta \neq -\alpha$, then

$$(0 \leq) \|T\|^2 - w^2(T) \leq \frac{1}{4} \frac{|\beta - \alpha|^2}{|\beta + \alpha|} (\|T\| + w(T)). \tag{2.192}$$

Proof. We give a short proof for the first inequality. The other results follow in a similar manner.

Utilizing the inequality (2.186) we can write that

$$\|Tx\|^2 \leq |\langle Tx, x \rangle|^2 + \frac{1}{4}|\beta - \alpha|^2 - \mathrm{Re}\langle C_{\alpha,\beta}(T)x, x \rangle, \tag{2.193}$$

for any $x \in H$, $\|x\| = 1$. Taking the supremum over $x \in H$, $\|x\| = 1$ in (2.193) we deduce the first inequality in (2.190). ∎

Remark 94. An equivalent and perhaps more useful version of (2.192) is the inequality

$$(w(T) \leq) \|T\| \leq \frac{1}{4} \cdot \frac{|\beta - \alpha|^2}{|\beta + \alpha|} + w(T),$$

provided that α and β satisfy the corresponding conditions mentioned in Theorem 93. Similar statements can be made for the other versions of this inequality presented below.

Corollary 95. *If $T \in B(H)$ and $M > m > 0$ are such that the transform $C_{m,M}(T) = (T^* - mI)(MI - T)$ is accretive, then*

$$(0 \leq) \|T\|^2 - w^2(T) \leq \begin{cases} \frac{1}{4}(M - m)^2 - \vartheta_i(C_{m,M}(T)), \\[2mm] \frac{1}{4}(M - m)^2 - w_i^2\left(T - \frac{m+M}{2}I\right), \\[2mm] \frac{1}{4}\frac{(M-m)^2}{mM}w^2(T), \\[2mm] \left(\sqrt{M} - \sqrt{m}\right)^2 w(T), \\[2mm] \frac{1}{4}\frac{(M-m)^2}{M+m}(\|T\| + w(T)). \end{cases} \tag{2.194}$$

The following result is well known in the literature (see for instance [49]):

$$w(T^n) \leq w^n(T),$$

for each positive integer n and any operator $T \in B(H)$.

The following reverse inequalities for $n = 2$ can be stated:

Theorem 96 (Dragomir [19], 2008). *Let $T \in B(H)$ and $\alpha, \beta \in \mathbb{K}$ be so that the transform $C_{\alpha,\beta}(T)$ is accretive, then*

$$(0 \leq) w^2(T) - w(T^2) \leq \frac{1}{4} |\beta - \alpha|^2 - \begin{cases} \vartheta_i\left(C_{\alpha,\beta}(T)\right) \\ w_i^2\left(T - \frac{\alpha+\beta}{2}I\right) \end{cases} \tag{2.195}$$

$$\left(\leq \frac{1}{4} |\beta - \alpha|^2 \right).$$

If $\operatorname{Re}(\beta\overline{\alpha}) > 0$, *then*

$$(0 \leq) w^2(T) - w(T^2) \leq \begin{cases} \frac{1}{4} \frac{|\beta-\alpha|^2}{\operatorname{Re}(\beta\overline{\alpha})} w^2(T), \\ \left(|\alpha + \beta| - 2\left[\operatorname{Re}(\beta\overline{\alpha})\right]^{\frac{1}{2}}\right) w(T). \end{cases} \tag{2.196}$$

If $\beta \neq -\alpha$, *then*

$$(0 \leq) w^2(T) - w(T^2) \leq \frac{1}{4} \frac{|\beta - \alpha|^2}{|\beta + \alpha|} (\|T\| + w(T)). \tag{2.197}$$

Proof. We give a short proof for the first inequality only. The other inequalities can be proved in a similar manner.

Utilizing the inequality (2.186) we can write that

$$|\langle Tx, x\rangle|^2 - |\langle T^2 x, x\rangle| \leq \left| \langle T^2 x, x\rangle - [\langle Tx, x\rangle]^2 \right|$$

$$\leq \frac{1}{4} |\beta - \alpha|^2 - \operatorname{Re}\langle C_{\alpha,\beta}(T) x, x\rangle,$$

for any $x \in H$, $\|x\| = 1$, which implies that

$$|\langle Tx, x\rangle|^2 \leq |\langle T^2 x, x\rangle| + \frac{1}{4} |\beta - \alpha|^2 - \operatorname{Re}\langle C_{\alpha,\beta}(T) x, x\rangle \tag{2.198}$$

for any $x \in H$, $\|x\| = 1$. Taking the supremum over $x \in H$, $\|x\| = 1$ in (2.198) we deduce the desired inequality in (2.195). ∎

Remark 97. If $T \in B(H)$ and $M > m > 0$ are such that the transform $C_{m,M}(T) = (T^* - mI)(MI - T)$ is accretive, then all the inequalities in (2.194) hold true with the left side replaced by the nonnegative quantity $w^2(T) - w(T^2)$.

2.5.5   New Inequalities of Kantorovich Type

The following result comprising some inequalities for the Kantorovich functional can be stated:

Theorem 98 (Dragomir [19], 2008). *Let $T \in B(H)$ and $\alpha, \beta \in \mathbb{K}$ be such that the transform $C_{\alpha,\beta}(T)$ is accretive. If $\operatorname{Re}(\beta\overline{\alpha}) > 0$ and the operator $-i \operatorname{Im}(\beta\overline{\alpha}) C_{\alpha,\beta}(T)$ is accretive, then*

$$|K(T;x) - 1| \tag{2.199}$$

$$\leq \begin{cases} \frac{1}{4}\frac{|\beta-\alpha|^2}{|\beta\alpha|} - \left[\operatorname{Re}\langle C_{\alpha,\beta}(T)x, x\rangle \operatorname{Re}\left\langle C_{\frac{1}{\alpha},\frac{1}{\beta}}\left((T^*)^{-1}\right)x, x\right\rangle\right]^{\frac{1}{2}}, \\[2mm] \frac{1}{4}\frac{|\beta-\alpha|^2}{|\beta\alpha|} - \left|\left\langle\left(T - \frac{\alpha+\beta}{2}I\right)x, x\right\rangle\right|\left|\left\langle\left(T^{-1} - \frac{\alpha+\beta}{2\alpha\beta}I\right)x, x\right\rangle\right|, \\[2mm] \frac{1}{4}\frac{|\beta-\alpha|^2}{\operatorname{Re}(\beta\overline{\alpha})}|K(T;x)|, \\[2mm] \frac{|\beta+\alpha| - 2[\operatorname{Re}(\beta\overline{\alpha})]^{\frac{1}{2}}}{|\beta\alpha|^{\frac{1}{2}}}|K(T;x)|^{\frac{1}{2}}, \\[2mm] \frac{1}{4}\frac{|\beta-\alpha|^2}{|\beta\alpha|^{\frac{1}{2}}|\beta+\alpha|}\left[(\|Tx\| + |\langle Tx, x\rangle|)\left(\left\|(T^*)^{-1}x\right\| + |\langle T^{-1}x, x\rangle|\right)\right]^{\frac{1}{2}}, \end{cases}$$

for any $x \in H$, $\|x\| = 1$.

Proof. Utilizing the identity (2.180), we have for each $x \in H$, $\|x\| = 1$ that

$$\operatorname{Re}\left\langle C_{\frac{1}{\alpha},\frac{1}{\beta}}\left(T^{-1}\right)x, x\right\rangle$$

$$= \frac{1}{|\beta\alpha|^2}\operatorname{Re}\left[\overline{\beta\overline{\alpha}}\left\langle\left(T^{-1}\right)^*C_{\alpha,\beta}(T)T^{-1}x, x\right\rangle\right]$$

$$= \frac{1}{|\beta\alpha|^2}\left[\operatorname{Re}(\beta\overline{\alpha}) \cdot \operatorname{Re}\left\langle\left(T^{-1}\right)^*C_{\alpha,\beta}(T)T^{-1}x, x\right\rangle\right.$$

$$\left. + \operatorname{Im}(\beta\overline{\alpha}) \cdot \operatorname{Im}\left\langle\left(T^{-1}\right)^*C_{\alpha,\beta}(T)T^{-1}x, x\right\rangle\right]$$

$$= \frac{1}{|\beta\alpha|^2}\left[\operatorname{Re}(\beta\overline{\alpha}) \cdot \operatorname{Re}\left\langle\left(T^{-1}\right)^*C_{\alpha,\beta}(T)T^{-1}x, x\right\rangle\right.$$

$$\left. + \operatorname{Re}\left\langle\left(T^{-1}\right)^*\left(-i\operatorname{Im}(\beta\overline{\alpha})C_{\alpha,\beta}(T)\right)T^{-1}x, x\right\rangle\right]$$

$$\geq 0,$$

showing that the operator $C_{\frac{1}{\alpha},\frac{1}{\beta}}\left(T^{-1}\right)$ is also accretive.

Now, on applying Theorem 88 for the difference $\langle BTx, x \rangle - \langle Tx, x \rangle \langle Bx, x \rangle$ and for the choices $B = T^{-1}$ and $\delta = 1/\beta$, $\gamma = 1/\alpha$, we get the desired inequality (2.199). The details are omitted.   ∎

Remark 99. A sufficient simple condition for the second assumption to hold in the above theorem is that $\beta\bar{\alpha}$ is a positive real number.

Remark 100. The third and the fourth inequalities in (2.199) can be written in the following equivalent forms that perhaps are more useful

$$\left| K^{-1}(T;x) - 1 \right| \le \frac{1}{4} \frac{|\beta - \alpha|^2}{\mathrm{Re}\,(\beta\bar{\alpha})}$$

and

$$\left| K^{1/2}(T;x) - K^{-1/2}(T;x) \right| \le \frac{|\beta + \alpha| - 2\,[\mathrm{Re}\,(\beta\bar{\alpha})]^{\frac{1}{2}}}{|\beta\alpha|^{\frac{1}{2}}},$$

provided that α and β satisfy the assumptions in Theorem 98. Similar comments apply for the other related results listed below.

However, for practical applications, the following even more particular case is of interest:

Corollary 101. *Let* $T \in B(H)$ *and* $M > m > 0$ *are such that the transform* $C_{m,M}(T) = (T^* - mI)(MI - T)$ *is accretive. Then*

$|K(T;x) - 1|$

$$\le \begin{cases} \frac{1}{4} \frac{(M-m)^2}{mM} - \left[\mathrm{Re}\,\langle C_{m,M}(T)x, x \rangle \, \mathrm{Re}\,\left\langle C_{\frac{1}{m},\frac{1}{M}}\left((T^*)^{-1}\right)x, x \right\rangle \right]^{\frac{1}{2}}, \\[2ex] \frac{1}{4} \frac{(M-m)^2}{mM} - \left| \left\langle \left(T - \frac{m+M}{2}I\right)x, x \right\rangle \right| \left| \left\langle \left(T^{-1} - \frac{m+M}{2mM}I\right)x, x \right\rangle \right|, \\[2ex] \frac{1}{4} \frac{(M-m)^2}{mM} \, |K(T;x)|, \\[2ex] \frac{(\sqrt{M} - \sqrt{m})^2}{\sqrt{mM}} \, |K(T;x)|^{\frac{1}{2}}, \\[2ex] \frac{1}{4} \frac{(M-m)^2}{\sqrt{mM}\,(m+M)} \left[\left(\|Tx\| + |\langle Tx, x \rangle| \right) \left(\left\| (T^*)^{-1}x \right\| + |\langle T^{-1}x, x \rangle| \right) \right]^{\frac{1}{2}}, \end{cases}$$

$$\tag{2.200}$$

for any $x \in H$, $\|x\| = 1$.

Finally, on returning to the original assumptions, we can state the following results which provide refinements for the additive version of the operator Kantorovich inequality (2.169) as well as other similar results that apparently are new:

Corollary 102. *Let T be a self-adjoint operator on H and $M > m > 0$ such that $MI \geq T \geq mI$ in the partial operator order of $B(H)$. Then*

$$0 \leq K(T;x) - 1$$

$$\leq \begin{cases} \frac{1}{4} \frac{(M-m)^2}{mM} - \left[\mathrm{Re} \left\langle C_{m,M}(T)x, x \right\rangle \mathrm{Re} \left\langle C_{\frac{1}{m},\frac{1}{M}}(T^{-1})x, x \right\rangle \right]^{\frac{1}{2}}, \\[2mm] \frac{1}{4} \frac{(M-m)^2}{mM} - \left| \left\langle (T - \frac{m+M}{2}I)x, x \right\rangle \right| \left| \left\langle (T^{-1} - \frac{m+M}{2mM}I)x, x \right\rangle \right|, \\[2mm] \frac{(\sqrt{M}-\sqrt{m})^2}{\sqrt{mM}} [K(T;x)]^{\frac{1}{2}}, \\[2mm] \frac{1}{4} \frac{(M-m)^2}{\sqrt{mM}(m+M)} \left[(\|Tx\| + \langle Tx, x \rangle)(\|T^{-1}x\| + \langle T^{-1}x, x \rangle) \right]^{\frac{1}{2}}, \end{cases} \qquad (2.201)$$

for any $x \in H$, $\|x\| = 1$.

The proof is obvious by Corollary 102 on noticing the fact that $MI \geq T \geq mI$ for a self-adjoint operator T implies that $C_{m,M}(T) = (T^* - mI)(MI - T)$ is accretive. The reverse is not true in general.

References

1. Braess, D.: Finite elements. Theory, Fast Solvers, and Applications in Solid Mechanics, 2nd edn., pp. xviii+352. Translated from the 1992 German edition by Larry L. Schumaker. Cambridge University Press, Cambridge (2001)
2. Buzano, M.L.: Generalizzazione della disiguaglianza di Cauchy-Schwaz (Italian). Rend. Sem. Mat. Univ. e Politech. Torino **31** (1971/73), 405–409 (1974)
3. Diaz, J.B., Metcalf, F.T.: Complementary inequalities. III. Inequalities complementary to Schwarz's inequality in Hilbert space. Math. Ann. **162**, 120–139 (1965/1966)
4. Dragomir, S.S.: Some refinements of Schwarz inequality, Simposional de Matematică şi Aplicaţii, Polytechnical Institute Timişoara, Romania, 1–2 Nov 1985, 13–16. ZBL 0594:46018.
5. Dragomir, S.S.: A generalisation of Grüss' inequality in inner product spaces and applications. J. Math. Anal. Appl. **237**, 74–82 (1999)
6. Dragomir, S.S.: Some Grüss type inequalities in inner product spaces. J. Ineq. Pure & Appl. Math. **4**(2), Art. 42 (2003),
7. Dragomir, S.S.: New reverses of Schwarz, triangle and Bessel inequalities in inner product spaces. Austral. J. Math. Anal. & Applics. **1**(1), Article 1 (2004)
8. Dragomir, S.S.: On Bessel and Grüss inequalities for orthornormal families in inner product spaces. Bull. Austral. Math. Soc. **69**(2), 327–340 (2004)
9. Dragomir, S.S.: Reverses of Schwarz, triangle and Bessel inequalities in inner product spaces. J. Inequal. Pure & Appl. Math. **5**(3), Article 76 (2004)
10. Dragomir, S.S.: Semi Inner Products and Applications, Nova Science Publishers Inc., New York (2004)

11. Dragomir, S.S.: Advances in Inequalities of the Schwarz, Grüss and Bessel Type in Inner Product Spaces, pp. x+249. Nova Science Publishers Inc, New York (2005)

12. Dragomir, S.S.: Inequalities for the norm and the numerical radius of composite operators in Hilbert spaces, Inequalities and applications. Internat. Ser. Numer. Math. 157, 135–146 Birkhäuser, Basel, 2009. Preprint available in RGMIA Res. Rep. Coll. **8**, Article 11 (2005), Supplement

13. Dragomir, S.S.: Reverse inequalities for the numerical radius of linear operators in Hilbert spaces. Bull. Austral. Math. Soc. **73**, 255–262 (2006). Preprint available on line at RGMIA Res. Rep. Coll. **8**, Article 9 (2005)

14. Dragomir, S.S.: A potpourri of Schwarz related inequalities in inner product spaces (II), J. Ineq. Pure & Appl. Math. **7**(1), Art. 14 (2006)

15. Dragomir, S.S.: Reverses of the Schwarz inequality in inner product spaces generalising a Klamkin-McLenaghan result. Bull. Austral. Math. Soc. **73**(1), 69–78 (2006)

16. Dragomir, S.S.: Inequalities for the norm and the numerical radius of linear operators in Hilbert spaces. Demonstratio Math. **40**(2), 411–417 (2007)

17. Dragomir, S.S.: Inequalities for some functionals associated with bounded linear operators in Hilbert spaces. Publ. Res. Inst. Math. Sci. **43**(4), 1095–1110 (2007)

18. Dragomir, S.S.: Inequalities for the numerical radius, the norm and the maximum of the real part of bounded linear operators in Hilbert spaces. Linear Algebra Appl. **428**(11–12), 2980–2994 (2008)

19. Dragomir, S.S.: New inequalities of the Kantorovich type for bounded linear operators in Hilbert spaces. Linear Algebra Appl. **428**(11–12), 2750–2760 (2008)

20. Dragomir, S.S., Sándor, J.: Some inequalities in prehilbertian spaces. Studia Univ. "Babeş-Bolyai"–Mathematica **32**(1), 71–78 (1987)

21. Fujii, M., Kamei, E., Matsumoto, A.: Parameterized Kantorovich inequality for positive operators. Nihonkai Math. J. **6**(2), 129–134 (1995)

22. Fujii, M., Izumino, S., Nakamoto, R., Seo, Y.: Operator inequalities related to Cauchy-Schwarz and Hölder-McCarthy inequalities. Nihonkai Math. J. **8**(2), 117–122 (1997)

23. Fuji, M., Nakamura, M.: Jensen inequality is a complement to Kantorovich inequality. Sci. Math. Jpn. **62**(1), 39–45 (2005)

24. Fujii, M., Nakamura, M.: Kadison's Schwarz inequality and noncommutative Kantorovich inequality. Sci. Math. Jpn. **63**(1), 101–102 (2006)

25. Furuta, T.: Extensions of Hölder-McCarthy and Kantorovich inequalities and their applications. Proc. Japan Acad. Ser. A Math. Sci. **73**(3), 38–41 (1997)

26. Furuta, T.: Operator inequalities associated with Hölder-McCarthy and Kantorovich inequalities. J. Inequal. Appl. **2**(2), 137–148 (1998)

27. Furuta, T., Giga, M.: A complementary result of Kantorovich type order preserving inequalities by Mičić-Pečarić-Seo. Linear Algebra Appl. **369**, 27–40 (2003)

28. Galantai, A.: A study of Auchmuty's error estimate. Numerical methods and computational mechanics (Miskolc, 1998). Comput. Math. Appl. **42**(8–9), 1093–1102 (2001)

29. Goldstein, A.A.: A modified Kantorovich inequality for the convergence of Newton's method. Mathematical developments arising from linear programming (Brunswick, ME, 1988), 285–294, Contemp. Math., 114, Amer. Math. Soc. Providence, RI (1990)

30. Goldstein, A.A.: A global Newton method. II. Analytic centers. Math. Programming **62**(2), Ser. B, 223–237 (1993)

31. Goldstein, A., Ryff, J.V., Clarke, L.E.: Problem 5473. Amer. Math. Monthly **75**(3), 309 (1968)

32. Greub, W., Rheinboldt, W.: On a generalization of an inequality of L. V. Kantorovich. Proc. Amer. Math. Soc. **10** 407–415 (1959)

33. Gustafson, K.E., Rao, D.K.M.: Numerical Range, Springer, New York, Inc. (1997)

34. Halmos, P.R.: A Hilbert Space Problem Book, 2nd edn. Springer, New York, Heidelberg, Berlin (1982)

35. Householder, A.S.: The Kantorovich and some related inequalities. SIAM Rev. **7**, 463–473 (1965)

36. Kantorovič, L.V.: Functional analysis and applied mathematics. (Russian) Uspehi Matem. Nauk (N.S.) **3** 6(28), 89–185 (1948)
37. Khatri, C.G., Rao, C.R.: Some extensions of the Kantorovich inequality and statistical applications. J. Multivariate Anal. **11**(4), 498–505 (1981)
38. Kittaneh, F.: A numerical radius inequality and an estimate for the numerical radius of the Frobenius companion matrix. Studia Math. **158**(1), 11–17 (2003)
39. Kittaneh, F.: Numerical radius inequalities for Hilbert space operators. Studia Math. **168**(1), 73–80 (2005)
40. Lin, C.T.: Extrema of quadratic forms and statistical applications. Comm. Statist. A—Theory Methods **13**(12), 1517–1520 (1984)
41. Liu, S., Neudecker, H.: Kantorovich inequalities and efficiency comparisons for several classes of estimators in linear models. Statist. Neerlandica **51**(3), 345–355 (1997)
42. Lumer, G.: Semi-inner-product spaces. Trans. Amer. Math. Soc. **100**(1), 29–43 (1961)
43. Merikoski, J.K., Kumar, P.: Lower bounds for the numerical radius. Lin. Alg. Appl. **410**, 135–142 (2005)
44. Mond, B.: An inequality for operators in a Hilbert space. Pacific J. Math. **18**, 161–163 (1966)
45. Mond, B., Pečarić, J.E.: Converses of Jensen's inequality for linear maps of operators. An. Univ. Timisoara Ser. Mat.-Inform. **31**(2), 223–228 (1993)
46. Mond, B., Pečarić, J.E.: Converses of Jensen's inequality for several operators. Rev. Anal. Numér. Théor. Approx. **23**(2), 179–183 (1994)
47. Mond, B., Shisha, O.: A difference inequality for operators in Hilbert space. Blanch Anniversary Volume pp. 269–275. Aerospace Research Lab, U.S. Air Force, Washington (1967)
48. Nakamoto, R., Nakamura, M.: Operator mean and Kantorovich inequality. Math. Japon. **44**(3), 495–498 (1996)
49. Pearcy, C.: An elementary proof of the power inequality for the numerical radius. Michigan Math. J. **13**, 289–291 (1966)
50. Pečarić, J.E., Puntanen, S. Styan, G.P.H.: Some further matrix extensions of the Cauchy-Schwarz and Kantorovich inequalities, with some statistical applications. Special issue honoring Calyampudi Radhakrishna Rao. Lin. Algebra Appl. **237/238**, 455–476 (1996)
51. Rao, C.R.: The inefficiency of least squares: extensions of the Kantorovich inequality. Linear Algebra Appl. **70**, 249–255 (1985)
52. Robinson, P.D., Wathen, A.J.: Variational bounds on the entries of the inverse of a matrix. IMA J. Numer. Anal. **12**(4), 463–486 (1992)
53. Söderlind, G.: The logarithmic norm. History and modern theory. BIT **46**, 631–652 (2006)
54. Spain, P.G.: Operator versions of the Kantorovich inequality. Proc. Amer. Math. Soc. **124**(9), 2813–2819 (1996)
55. Spall, J.C.: The Kantorovich inequality for error analysis of the Kalman filter with unknown noise distributions. Automatica J. IFAC **31**(10), 1513–1517. (1995)
56. Styan, G.P.H.: On some inequalities associated with ordinary least squares and the Kantorovich inequality. Acta Univ. Tamper. Ser. A **153**, 158–166 (1983)
57. Strang, G.: On the Kantorovich inequality. Proc. Amer. Math. Soc. **11**, 468 (1960)
58. Tsukada, M., Takahasi, S.-E.: The best possibility of the bound for the Kantorovich inequality and some remarks. J. Inequal. Appl. **1**(4), 327–334 (1997)
59. Watson, G.S.: A method for discovering Kantorovich-type inequalities and a probabilistic interpretation. Linear Algebra Appl **97**, 211–217 (1987)
60. Watson, G.S., Alpargu, G., Styan, G.P.H.: Some comments on six inequalities associated with the inefficiency of ordinary least squares with one regressor. Lin. Algebra Appl. **264**, 13–54 (1997)
61. Yamazaki, T.: An extension of Specht's theorem via Kantorovich inequality and related results. Math. Inequal. Appl. **3**(1), 89–96 (2000)
62. Yang, H.: Efficiency matrix and the partial ordering of estimate. Comm. Statist. Theory Methods **25**(2), 457–468 (1996)

Chapter 3
Inequalities for Two Operators

In this chapter we present recent results obtained by the author concerning the norms and the numerical radii of two bounded linear operators. The proofs of the results are elementary. Some vector inequalities in inner product spaces as well as inequalities for means of nonnegative real numbers are also employed.

3.1 General Inequalities for Numerical Radius

3.1.1 Preliminary Facts

The following result may be stated:

Theorem 103 (Dragomir [7], 2005). *Let $A, B : H \to H$ be two bounded linear operators on the Hilbert space $(H, \langle \cdot, \cdot \rangle)$. If $r > 0$ and*

$$\|A - B\| \leq r, \tag{3.1}$$

then

$$\left\| \frac{A^*A + B^*B}{2} \right\| \leq w\left(B^*A\right) + \frac{1}{2}r^2. \tag{3.2}$$

Proof. For any $x \in H$, $\|x\| = 1$, we have from (3.1) that

$$\|Ax\|^2 + \|Bx\|^2 \leq 2\,\mathrm{Re}\,\langle Ax, Bx \rangle + r^2. \tag{3.3}$$

However

$$\|Ax\|^2 + \|Bx\|^2 = \left\langle \left(A^*A + B^*B\right)x, x \right\rangle$$

S.S. Dragomir, *Inequalities for the Numerical Radius of Linear Operators in Hilbert Spaces*, SpringerBriefs in Mathematics, DOI 10.1007/978-3-319-01448-7_3, © Silvestru Sever Dragomir 2013

and by (3.3) we obtain

$$\langle (A^*A + B^*B) x, x \rangle \leq 2 \left| \langle (B^*A) x, x \rangle \right| + r^2 \tag{3.4}$$

for any $x \in H$, $\|x\| = 1$.

Taking the supremum over $x \in H$, $\|x\| = 1$ in (3.4) we get

$$w \left(A^*A + B^*B \right) \leq 2w \left(B^*A \right) + r^2 \tag{3.5}$$

and since the operator $A^*A + B^*B$ is self-adjoint, hence $w \left(A^*A + B^*B \right) = \|A^*A + B^*B\|$ and by (3.5) we deduce the desired inequality (3.2). ∎

Remark 104. We observe that, from the proof of the above theorem, we have the inequalities

$$0 \leq \left\| \frac{A^*A + B^*B}{2} \right\| - w \left(B^*A \right) \leq \frac{1}{2} \|A - B\|^2 , \tag{3.6}$$

provided that A, B are bounded linear operators in H.

The second inequality in (3.6) is obvious while the first inequality follows by the fact that

$$\langle (A^*A + B^*B) x, x \rangle = \|Ax\|^2 + \|Bx\|^2 \geq 2 \left| \langle (B^*A) x, x \rangle \right|$$

for any $x \in H$.

The inequality (3.2) is obviously a rich source of particular inequalities of interest.

Indeed, if we assume, for $\lambda \in \mathbb{C}$ and a bounded linear operator T, that we have $\|T - \lambda T^*\| \leq r$, for a given positive number r, then by (3.6) we deduce the inequality

$$0 \leq \left\| \frac{T^*T + |\lambda|^2 \, T T^*}{2} \right\| - |\lambda| \, w \left(T^2 \right) \leq \frac{1}{2} r^2. \tag{3.7}$$

Now, if we assume that for $\lambda \in \mathbb{C}$ and a bounded linear operator V we have that $\|V - \lambda I\| \leq r$, where I is the identity operator on H, then by (3.2) we deduce the inequality

$$0 \leq \left\| \frac{V^*V + |\lambda|^2 \, I}{2} \right\| - |\lambda| \, w \left(V \right) \leq \frac{1}{2} r^2.$$

As a dual approach, the following result may be noted as well:

Theorem 105 (Dragomir [7], 2005). *Let $A, B : H \to H$ be two bounded linear operators on the Hilbert space H. Then*

$$\left\| \frac{A + B}{2} \right\|^2 \leq \frac{1}{2} \left[\left\| \frac{A^*A + B^*B}{2} \right\| + w \left(B^*A \right) \right]. \tag{3.8}$$

Proof. We obviously have

$$\|Ax + Bx\|^2 = \|Ax\|^2 + 2\operatorname{Re}\langle Ax, Bx\rangle + \|Bx\|^2$$
$$\leq \langle (A^*A + B^*B)x, x\rangle + 2\left|\langle (B^*A)x, x\rangle\right|$$

for any $x \in H$.

Taking the supremum over $x \in H$, $\|x\| = 1$, we get $\|A + B\|^2 \leq \|A^*A + B^*B\| + 2w(B^*A)$, from where we get the desired inequality (3.8). ∎

Remark 106. The inequality (3.8) can generate some interesting particular results such as the following inequality:

$$\left\|\frac{T + T^*}{2}\right\|^2 \leq \frac{1}{2}\left[\left\|\frac{T^*T + TT^*}{2}\right\| + w\left(T^2\right)\right], \tag{3.9}$$

holding for each bounded linear operator $T : H \to H$.

The following result may be stated as well.

Theorem 107 (Dragomir [7], 2005). *Let $A, B : H \to H$ be two bounded linear operators on the Hilbert space H and $p \geq 2$. Then*

$$\left\|\frac{A^*A + B^*B}{2}\right\|^{\frac{p}{2}} \leq \frac{1}{4}\left[\|A - B\|^p + \|A + B\|^p\right]. \tag{3.10}$$

Proof. We use the following inequality for vectors in inner product spaces obtained by Dragomir and Sándor in [18]:

$$2\left(\|a\|^p + \|b\|^p\right) \leq \|a + b\|^p + \|a - b\|^p \tag{3.11}$$

for any $a, b \in H$ and $p \geq 2$.

Utilizing (3.11) we may write

$$2\left(\|Ax\|^p + \|Bx\|^p\right) \leq \|Ax + Bx\|^p + \|Ax - Bx\|^p \tag{3.12}$$

for any $x \in H$.

Now, observe that $\|Ax\|^p + \|Bx\|^p = \left(\|Ax\|^2\right)^{\frac{p}{2}} + \left(\|Bx\|^2\right)^{\frac{p}{2}}$ and by the elementary inequality $\frac{\alpha^q + \beta^q}{2} \geq \left(\frac{\alpha + \beta}{2}\right)^q$, $\alpha, \beta \geq 0$ and $q \geq 1$ we have

$$\left(\|Ax\|^2\right)^{\frac{p}{2}} + \left(\|Bx\|^2\right)^{\frac{p}{2}} \geq 2^{1-\frac{p}{2}}\left[\langle (A^*A + B^*B)x, x\rangle\right]^{\frac{p}{2}}. \tag{3.13}$$

Combining (3.12) with (3.13) we get

$$\frac{1}{4}\left[\|Ax - Bx\|^p + \|Ax + Bx\|^p\right] \geq \left|\left\langle\left(\frac{A^*A + B^*B}{2}\right)x, x\right\rangle\right|^{\frac{p}{2}} \tag{3.14}$$

for any $x \in H$, $\|x\| = 1$. Taking the supremum over $x \in H$, $\|x\| = 1$, and taking into account that $w\left(\frac{A^*A+B^*B}{2}\right) = \left\|\frac{A^*A+B^*B}{2}\right\|$, we deduce the desired result (3.10).   ∎

Remark 108. If $p = 2$, then we have the inequality:

$$\left\|\frac{A^*A + B^*B}{2}\right\| \le \left\|\frac{A - B}{2}\right\|^2 + \left\|\frac{A + B}{2}\right\|^2,$$

for any A, B bounded linear operators. This result can be obtained directly on utilizing the parallelogram identity as well. We also should observe that for $A = T$ and $B = T^*$, T a normal operator, the inequality (3.10) becomes

$$\|T\|^p \le \frac{1}{4}\left[\|T - T^*\|^p + \|T + T^*\|^p\right],$$

where $p \ge 2$.

The following result may be stated as well.

Theorem 109 (Dragomir [7], 2005). *Let $A, B : H \to H$ be two bounded linear operators on the Hilbert space H and $r \ge 1$. If $A^*A \ge B^*B$ in the operator order or, equivalently, $\|Ax\| \ge \|Bx\|$ for any $x \in H$, then*

$$\left\|\frac{A^*A + B^*B}{2}\right\|^r$$

$$\le \|A\|^{r-1}\|B\|^{r-1}w\left(B^*A\right) + \frac{1}{2}r^2\|A\|^{2r-2}\|A - B\|^2. \tag{3.15}$$

Proof. We use the following inequality for vectors in inner product spaces due to Goldstein, Ryff and Clarke [19]:

$$\|a\|^{2r} + \|b\|^{2r} \le 2\|a\|^{r-1}\|b\|^{r-1}\operatorname{Re}\langle a, b\rangle + r^2\|a\|^{2r-2}\|a - b\|^2, \tag{3.16}$$

where $r \ge 1$, $a, b \in H$ and $\|a\| \ge \|b\|$.

Utilizing (3.16) we can state that:

$$\|Ax\|^{2r} + \|Bx\|^{2r}$$

$$\le 2\|Ax\|^{r-1}\|Bx\|^{r-1}|\langle Ax, Bx\rangle| + r^2\|Ax\|^{2r-2}\|Ax - Bx\|^2, \tag{3.17}$$

for any $x \in H$. As in the proof of Theorem 107, we also have

$$2^{1-r}\left[\langle(A^*A + B^*B)x, x\rangle\right]^r \le \|Ax\|^{2r} + \|Bx\|^{2r}, \tag{3.18}$$

for any $x \in H$. Therefore, by (3.17) and (3.18), we deduce

$$\left[\left\langle\left(\frac{A^*A + B^*B}{2}\right)x, x\right\rangle\right]^r$$

$$\leq \|Ax\|^{r-1}\|Bx\|^{r-1}|\langle Ax, Bx\rangle| + \frac{1}{2}r^2\|A\|^{2r-2}\|Ax - Bx\|^2 \tag{3.19}$$

for any $x \in H$.

Taking the supremum in (3.19) we obtain the desired result (3.15). ∎

Remark 110. Following [20, p. 156], we recall that the bounded linear operator V is hyponormal, if $\|V^*x\| \leq \|Vx\|$ for all $x \in H$. Now, if we choose in (3.15) $A = V$ and $B = V^*$, then, on taking into account that for hyponormal operators $w(V^2) = \|V\|^2$, we get the inequality

$$\left\|\frac{V^*V + VV^*}{2}\right\|^r \leq \|V\|^{2r-2}\left[\|V\|^2 + \frac{1}{2}r^2\|V - V^*\|^2\right], \tag{3.20}$$

holding for any hyponormal operator V and any $r \geq 1$.

3.1.2 Further Inequalities for an Invertible Operator

In this section we assume that $B : H \to H$ is an invertible bounded linear operator and let $B^{-1} : H \to H$ be its inverse. Then, obviously,

$$\|Bx\| \geq \frac{1}{\|B^{-1}\|}\|x\| \quad \text{for any } x \in H, \tag{3.21}$$

where $\|B^{-1}\|$ denotes the norm of the inverse B^{-1}.

Theorem 111 (Dragomir [7], 2005). *Let $A, B : H \to H$ be two bounded linear operators on H and B is invertible such that, for a given $r > 0$,*

$$\|A - B\| \leq r. \tag{3.22}$$

Then

$$\|A\| \leq \|B^{-1}\|\left[w(B^*A) + \frac{1}{2}r^2\right]. \tag{3.23}$$

Proof. The condition (3.22) is obviously equivalent to:

$$\|Ax\|^2 + \|Bx\|^2 \leq 2\operatorname{Re}\langle(B^*A)x, x\rangle + r^2 \tag{3.24}$$

for any $x \in H$, $\|x\| = 1$. Since, by (3.21), $\|Bx\|^2 \geq \frac{1}{\|B^{-1}\|^2}\|x\|^2$, $x \in H$ and

$$\operatorname{Re}\langle(B^*A)x, x\rangle \leq |\langle(B^*A)x, x\rangle|,$$

hence by (3.24) we get

$$\|Ax\|^2 + \frac{\|x\|^2}{\|B^{-1}\|^2} \le 2\left|\langle (B^*A) x, x\rangle\right| + r^2 \qquad (3.25)$$

for any $x \in H$, $\|x\| = 1$. Taking the supremum over $x \in H$, $\|x\| = 1$ in (3.25), we have

$$\|A\|^2 + \frac{1}{\|B^{-1}\|^2} \le 2w\left(B^*A\right) + r^2. \qquad (3.26)$$

By the elementary inequality

$$\frac{2\|A\|}{\|B^{-1}\|} \le \|A\|^2 + \frac{1}{\|B^{-1}\|^2} \qquad (3.27)$$

and by (3.26) we then deduce the desired result (3.23).   ∎

Remark 112. If we choose above $B = \lambda I$, $\lambda \ne 0$, then we get the inequality

$$(0 \le) \|A\| - w(A) \le \frac{1}{2|\lambda|} r^2, \qquad (3.28)$$

provided $\|A - \lambda I\| \le r$. This result has been obtained in the earlier paper [9].

Also, if we assume that $B = \lambda A^*$, A is invertible, then we obtain

$$\|A\| \le \|A^{-1}\| \left[w\left(A^2\right) + \frac{1}{2|\lambda|} r^2 \right], \qquad (3.29)$$

provided $\|A - \lambda A^*\| \le r$, $\lambda \ne 0$.

The following result may be stated as well:

Theorem 113 (Dragomir [7], 2005). *Let $A, B : H \to H$ be two bounded linear operators on H. If B is invertible and for $r > 0$,*

$$\|A - B\| \le r, \qquad (3.30)$$

then

$$(0 \le) \|A\| \|B\| - w\left(B^*A\right) \le \frac{1}{2} r^2 + \frac{\|B\|^2 \|B^{-1}\|^2 - 1}{2\|B^{-1}\|^2}. \qquad (3.31)$$

Proof. The condition (3.30) is obviously equivalent to

$$\|Ax\|^2 + \|Bx\|^2 \le 2\,\mathrm{Re}\,\langle Ax, Bx\rangle + r^2$$

for any $x \in H$, which is clearly equivalent to

$$\|Ax\|^2 + \|B\|^2 \leq 2\operatorname{Re}\langle B^*Ax, x\rangle + r^2 + \|B\|^2 - \|Bx\|^2. \tag{3.32}$$

Since

$$\operatorname{Re}\langle B^*Ax, x\rangle \leq |\langle B^*Ax, x\rangle|, \quad \|Bx\|^2 \geq \frac{1}{\|B^{-1}\|^2}\|x\|^2$$

and

$$\|Ax\|^2 + \|B\|^2 \geq 2\|B\|\,\|Ax\|$$

for any $x \in H$, hence by (3.32), we get

$$2\|B\|\,\|Ax\| \leq 2|\langle B^*Ax, x\rangle| + r^2 + \frac{\|B\|^2\,\|B^{-1}\|^2 - 1}{\|B^{-1}\|^2} \tag{3.33}$$

for any $x \in H$, $\|x\| = 1$. Taking the supremum over $x \in H$, $\|x\| = 1$ we deduce the desired result (3.31). ∎

Remark 114. If we choose in Theorem 113, $B = \lambda A^*$, $\lambda \neq 0$, A is invertible, then we get the inequality:

$$(0 \leq)\|A\|^2 - w(A^2) \leq \frac{1}{2|\lambda|}r^2 + |\lambda| \cdot \frac{\|A\|^2\,\|A^{-1}\|^2 - 1}{2\|A^{-1}\|^2} \tag{3.34}$$

provided $\|A - \lambda A^*\| \leq r$.

The following result may be stated as well.

Theorem 115 (Dragomir [7], 2005). *Let $A, B : H \to H$ be two bounded linear operators on H. If B is invertible and for $r > 0$, we have*

$$\|A - B\| \leq r < \|B\|, \tag{3.35}$$

then

$$\|A\| \leq \frac{1}{\sqrt{\|B\|^2 - r^2}}\left(w(B^*A) + \frac{\|B\|^2\,\|B^{-1}\|^2 - 1}{2\|B^{-1}\|^2}\right). \tag{3.36}$$

Proof. The first part of condition (3.35) is obviously equivalent to

$$\|Ax\|^2 + \|Bx\|^2 \leq 2\operatorname{Re}\langle Ax, Bx\rangle + r^2$$

for any $x \in H$, which is clearly equivalent to

$$\|Ax\|^2 + \|B\|^2 - r^2 \leq 2 \operatorname{Re} \langle B^* Ax, x \rangle + \|B\|^2 - \|Bx\|^2 . \qquad (3.37)$$

Since

$$\operatorname{Re} \langle B^* Ax, x \rangle \leq |\langle B^* Ax, x \rangle| , \|Bx\|^2 \geq \frac{1}{\|B^{-1}\|^2} \|x\|^2$$

and, by the second part of (3.35),

$$\|Ax\|^2 + \|B\|^2 - r^2 \geq 2 \sqrt{\|B\|^2 - r^2} \, \|Ax\| ,$$

for any $x \in H$, hence by (3.37), we get

$$2 \|Ax\| \sqrt{\|B\|^2 - r^2} \leq 2 |\langle B^* Ax, x \rangle| + \frac{\|B\|^2 \|B^{-1}\|^2 - 1}{\|B^{-1}\|^2} \qquad (3.38)$$

for any $x \in H$, $\|x\| = 1$. Taking the supremum over $x \in H$, $\|x\| = 1$ in (3.38), we deduce the desired inequality (3.36). ∎

Remark 116. The above Theorem 115 has some particular cases of interest. For instance, if we choose $B = \lambda I$, with $|\lambda| > r$, then (3.35) is obviously fulfilled and by (3.36) we get

$$\|A\| \leq \frac{w(A)}{\sqrt{1 - \left(\frac{r}{|\lambda|} \right)^2}}, \qquad (3.39)$$

provided $\|A - \lambda I\| \leq r$. This result has been obtained in the earlier paper [9].

On the other hand, if in the above we choose $B = \lambda A^*$ with $\|A\| \geq \frac{r}{|\lambda|}$ $(\lambda \neq 0)$, then by (3.36) we get

$$\|A\| \leq \frac{1}{\sqrt{\|A\|^2 - \left(\frac{r}{|\lambda|} \right)^2}} \left[w(A^2) + |\lambda| \cdot \frac{\|A\|^2 \|A^{-1}\|^2 - 1}{2 \|A^{-1}\|^2} \right], \qquad (3.40)$$

provided $\|A - \lambda A^*\| \leq r$.

The following result may be stated as well.

Theorem 117 (Dragomir [7], 2005). *Let A, B and r be as in Theorem 111. Moreover, if*

$$\|B^{-1}\| < \frac{1}{r}, \qquad (3.41)$$

then

$$\|A\| \leq \frac{\|B^{-1}\|}{\sqrt{1 - r^2 \|B^{-1}\|^2}} w\left(B^* A\right).$$ (3.42)

Proof. Observe that, by (2.21) we have

$$\|A\|^2 + \frac{1 - r^2 \|B^{-1}\|^2}{\|B^{-1}\|^2} \leq 2w\left(B^* A\right).$$ (3.43)

Utilizing the elementary inequality

$$2\frac{\|A\|}{\|B^{-1}\|}\sqrt{1 - r^2 \|B^{-1}\|^2} \leq \|A\|^2 + \frac{1 - r^2 \|B^{-1}\|^2}{\|B^{-1}\|^2},$$ (3.44)

which can be stated since (3.41) is assumed to be true, hence by (3.43) and (3.44) we deduce the desired result (3.42). ∎

Remark 118. If we assume that $B = \lambda A^*$ with $\lambda \neq 0$ and A an invertible operator, then, by applying Theorem 117, we get the inequality:

$$\|A\| \leq \frac{\|A^{-1}\| w\left(A^2\right)}{\sqrt{|\lambda|^2 - r^2 \|A^{-1}\|^2}},$$ (3.45)

provided $\|A - \lambda A^*\| \leq r$ and $\|A^{-1}\| \leq \frac{|\lambda|}{r}$.

The following result may be stated as well.

Theorem 119 (Dragomir [7], 2005). *Let $A, B : H \to H$ be two bounded linear operators. If $r > 0$ and B is invertible with the property that $\|A - B\| \leq r$ and*

$$\frac{1}{\sqrt{r^2 + 1}} \leq \|B^{-1}\| < \frac{1}{r},$$ (3.46)

then

$$\|A\|^2 \leq w^2\left(B^* A\right) + 2w\left(B^* A\right) \cdot \frac{\|B^{-1}\| - \sqrt{1 - r^2 \|B^{-1}\|^2}}{\|B^{-1}\|}.$$ (3.47)

Proof. Let $x \in H$, $\|x\| = 1$. Then by (2.20) we have

$$\|Ax\|^2 + \frac{1}{\|B^{-1}\|^2} \leq 2\left|\langle B^* Ax, x\rangle\right| + r^2,$$ (3.48)

and since $\frac{1}{\|B^{-1}\|^2} - r^2 > 0$, we can conclude that $\left|\langle B^* Ax, x\rangle\right| > 0$ for any $x \in H$, $\|x\| = 1$.

Dividing (3.48) throughout by $|\langle B^*Ax, x\rangle| > 0$, we obtain

$$\frac{\|Ax\|^2}{|\langle B^*Ax, x\rangle|} \le 2 + \frac{r^2}{|\langle B^*Ax, x\rangle|} - \frac{1}{\|B^{-1}\|^2 \, |\langle B^*Ax, x\rangle|}. \tag{3.49}$$

Subtracting $|\langle B^*Ax, x\rangle|$ from both sides of (3.49), we get

$$\frac{\|Ax\|^2}{|\langle B^*Ax, x\rangle|} - |\langle B^*Ax, x\rangle|$$

$$\le 2 - |\langle B^*Ax, x\rangle| - \frac{1 - r^2 \, \|B^{-1}\|^2}{|\langle B^*Ax, x\rangle| \, \|B^{-1}\|^2}$$

$$= 2 - \frac{2\sqrt{1 - r^2 \, \|B^{-1}\|^2}}{\|B^{-1}\|}$$

$$- \left(\sqrt{|\langle B^*Ax, x\rangle|} - \frac{\sqrt{1 - r^2 \, \|B^{-1}\|^2}}{\|B^{-1}\| \, \sqrt{|\langle B^*Ax, x\rangle|}} \right)^2$$

$$\le 2 \left(\frac{\|B^{-1}\| - \sqrt{1 - r^2 \, \|B^{-1}\|^2}}{\|B^{-1}\|} \right), \tag{3.50}$$

which gives:

$$\|Ax\|^2 \le |\langle B^*Ax, x\rangle|^2$$

$$+ 2 \, |\langle B^*Ax, x\rangle| \, \frac{\|B^{-1}\| - \sqrt{1 - r^2 \, \|B^{-1}\|^2}}{\|B^{-1}\|}. \tag{3.51}$$

We also remark that, by (3.46) the quantity

$$\|B^{-1}\| - \sqrt{1 - r^2 \, \|B^{-1}\|^2} \ge 0,$$

hence, on taking the supremum in (3.51) over $x \in H$, $\|x\| = 1$, we deduce the desired inequality. ∎

Remark 120. It is interesting to remark that if we assume $\lambda \in \mathbb{C}$ with $0 < r \le |\lambda| \le \sqrt{r^2 + 1}$ and $\|A - \lambda I\| \le r$, then by (2.17) we can state the following inequality:

$$\|A\|^2 \le |\lambda|^2 \, w\left(A^2\right) + 2 \, |\lambda| \left(1 - \sqrt{|\lambda|^2 - r^2}\right) w\left(A\right). \tag{3.52}$$

Also, if $\|A - A^*\| \le r$, A is invertible and $\frac{1}{\sqrt{r^2+1}} \le \|A^{-1}\| \le \frac{1}{r}$, Then, by (3.47), we also have

$$\|A\|^2 \le w^2\left(A^2\right) + 2w\left(A^2\right) \cdot \frac{\|A^{-1}\| - \sqrt{1 - r^2\|A^{-1}\|^2}}{\|A^{-1}\|}. \tag{3.53}$$

One can also prove the following result.

Theorem 121. *Let $A, B : H \to H$ be two bounded linear operators. If $r > 0$ and B is invertible with the property that $\|A - B\| \le r$ and $\left\|B^{-1}\right\| \le \frac{1}{r}$, then*

$$(0 \le) \|A\|^2 \|B\|^2 - w^2\left(B^*A\right) \tag{3.54}$$

$$\le 2w\left(B^*A\right) \cdot \frac{\|B\|}{\|B^{-1}\|} \left(\|B\|\,\|B^{-1}\| - \sqrt{1 - r^2\|B^{-1}\|^2}\right).$$

Proof. We subtract the quantity $\frac{|\langle B^*Ax, x\rangle|}{\|B\|^2}$ from both sides of (3.49) to obtain

$$0 \le \frac{\|Ax\|^2}{|\langle B^*Ax, x\rangle|} - \frac{|\langle B^*Ax, x\rangle|}{\|B\|^2}$$

$$\le 2 - 2 \cdot \frac{\sqrt{1 - r^2\|B^{-1}\|^2}}{\|B\|\,\|B^{-1}\|}$$

$$- \left(\frac{\sqrt{|\langle B^*Ax, x\rangle|}}{\|B\|} - \frac{\sqrt{1 - r^2\|B^{-1}\|^2}}{\sqrt{|\langle B^*Ax, x\rangle|}\,\|B^{-1}\|}\right)^2 \tag{3.55}$$

$$\le 2 \cdot \frac{\left(\|B\|\,\|B^{-1}\| - \sqrt{1 - r^2\|B^{-1}\|^2}\right)}{\|B\|\,\|B^{-1}\|},$$

which is equivalent with

$$(0 \le) \|Ax\|^2 \|B\|^2 - |\langle B^*Ax, x\rangle|^2 \tag{3.56}$$

$$\le 2\frac{\|B\|}{\|B^{-1}\|} |\langle B^*Ax, x\rangle| \left(\|B\|\,\|B^{-1}\| - \sqrt{1 - r^2\|B^{-1}\|^2}\right)$$

for any $x \in H$, $\|x\| = 1$.

The inequality (3.56) also shows that $\|B\|\,\|B^{-1}\| \ge \sqrt{1 - r^2\|B^{-1}\|^2}$ and then, by (3.56), we get

$$\|Ax\|^2 \|B\|^2 \le |\langle B^* Ax, x \rangle|^2$$

$$+ 2 \frac{\|B\|}{\|B^{-1}\|} |\langle B^* Ax, x \rangle| \left(\|B\| \|B^{-1}\| - \sqrt{1 - r^2 \|B^{-1}\|^2} \right)$$

$$(3.57)$$

for any $x \in X$, $\|x\| = 1$. Taking the supremum in (3.57) we deduce the desired inequality (3.54). ∎

Remark 122. The above Theorem 121 has some particular instances of interest as follows. If, for instance, we choose $B = \lambda I$ with $|\lambda| \ge r > 0$ and $\|A - \lambda I\| \le r$, then by (3.54) we obtain the inequality

$$(0 \le) \|A\|^2 - w^2(A) \le 2 |\lambda| w(A) \left(1 - \sqrt{1 - \frac{r^2}{|\lambda|^2}} \right). \tag{3.58}$$

Also, if A is invertible, $\|A - \lambda A^*\| \le r$ and $\|A^{-1}\| \le \frac{|\lambda|}{r}$, then by (3.54) we can state:

$$(0 \le) \|A\|^4 - w^2(A^2)$$

$$\le 2 |\lambda| w(A^2) \cdot \frac{\|A\|}{\|A^{-1}\|} \left(\|A\| \|A^{-1}\| - \sqrt{1 - \frac{r^2}{|\lambda|^2} \|A^{-1}\|^2} \right). \tag{3.59}$$

3.2   Other Norm and Numerical Radius Inequalities

3.2.1   Other Norm and Numerical Radius Inequalities

For the complex numbers α, β and the bounded linear operator T we define the following transform (see [13]):

$$C_{\alpha,\beta}(T) := (T^* - \overline{\alpha} I)(\beta I - T), \tag{3.60}$$

where by T^* we denote the adjoint of T.

In light of the above results it is then natural to compare the quantities $\|AB\|$ and $w(A) w(B) + w(A) \|B\| + \|A\| w(B)$ provided that some information about the transforms $C_{\alpha,\beta}(A)$ and $C_{\gamma,\delta}(B)$ are available, where $\alpha, \beta, \gamma, \delta \in \mathbb{K}$.

Theorem 123 (Dragomir [12], 2009). *Let $A, B \in B(H)$ and $\alpha, \beta, \gamma, \delta \in \mathbb{K}$ be such that the transforms $C_{\alpha,\beta}(A)$ and $C_{\gamma,\delta}(B)$ are accretive, then*

$$\|BA\| \le w(A) w(B) + w(A) \|B\| + \|A\| w(B) + \frac{1}{4} |\beta - \alpha| |\gamma - \delta|. \tag{3.61}$$

Proof. Since $C_{\alpha,\beta}(A)$ and $C_{\gamma,\delta}(B)$ are accretive, then we have that

$$\left\| Ax - \frac{\alpha+\beta}{2}x \right\| \leq \frac{1}{2}|\beta-\alpha| \quad \text{and} \quad \left\| B^*x - \frac{\bar{\gamma}+\bar{\delta}}{2}x \right\| \leq \frac{1}{2}\left|\bar{\gamma}-\bar{\delta}\right|,$$

for any $x \in H$, $\|x\| = 1$.

Utilizing the Schwarz inequality we may write that

$$|\langle Ax - \langle Ax,x\rangle x, B^*y - \langle B^*y,y\rangle y\rangle|$$
$$\leq \|Ax - \langle Ax,x\rangle x\| \, \|B^*y - \langle B^*y,y\rangle y\|, \qquad (3.62)$$

for any $x, y \in H$, with $\|x\| = \|y\| = 1$.

Since for any vectors $u, f \in H$ with $\|f\| = 1$ we have $\|u - \langle u,f\rangle f\| = \inf_{\mu\in\mathbb{K}}\|u - \mu f\|$, then obviously

$$\|Ax - \langle Ax,x\rangle x\| \leq \left\| Ax - \frac{\alpha+\beta}{2}x \right\| \leq \frac{1}{2}|\beta-\alpha|$$

and

$$\|B^*y - \langle B^*y,y\rangle y\| \leq \left\| B^*y - \frac{\bar{\gamma}+\bar{\delta}}{2}y \right\| \leq \frac{1}{2}|\gamma-\delta|$$

producing the inequality

$$\|Ax - \langle Ax,x\rangle x\| \, \|B^*y - \langle B^*y,y\rangle y\| \leq \frac{1}{4}|\beta-\alpha|\,|\gamma-\delta|. \qquad (3.63)$$

Now, observe that

$$\langle Ax - \langle Ax,x\rangle x, B^*y - \langle B^*y,y\rangle y\rangle$$
$$= \langle BAx,y\rangle + \langle Ax,x\rangle\langle By,y\rangle\langle x,y\rangle - \langle Ax,x\rangle\langle Bx,y\rangle - \langle Ax,y\rangle\langle By,y\rangle,$$

for any $x, y \in H$, with $\|x\| = \|y\| = 1$.

Taking the modulus in the equality and utilizing its properties we have successively

$$|\langle Ax - \langle Ax,x\rangle x, B^*y - \langle B^*y,y\rangle y\rangle|$$
$$\geq |\langle BAx,y\rangle| - |\langle Ax,x\rangle\langle Bx,y\rangle + \langle Ax,y\rangle\langle By,y\rangle - \langle Ax,x\rangle\langle By,y\rangle\langle x,y\rangle|$$
$$\geq |\langle BAx,y\rangle| - |\langle Ax,x\rangle\langle Bx,y\rangle|$$
$$\quad - |\langle Ax,y\rangle\langle By,y\rangle| - |\langle Ax,x\rangle\langle By,y\rangle\langle x,y\rangle|$$

which is equivalent with

$$|\langle Ax - \langle Ax, x \rangle\, x,\, B^* y - \langle B^* y, y \rangle\, y \rangle|$$
$$+ |\langle Ax, x \rangle \langle Bx, y \rangle| + |\langle Ax, y \rangle \langle By, y \rangle| + |\langle Ax, x \rangle \langle By, y \rangle \langle x, y \rangle| \quad (3.64)$$
$$\geq |\langle BAx, y \rangle|,$$

for any $x, y \in H$, with $\|x\| = \|y\| = 1$.

Finally, on making use of the inequalities (3.62)–(3.64), we can state that

$$\frac{1}{4}|\beta - \alpha|\,|\gamma - \delta|$$
$$+ |\langle Ax, x \rangle \langle Bx, y \rangle| + |\langle Ax, y \rangle \langle By, y \rangle| + |\langle Ax, x \rangle \langle By, y \rangle \langle x, y \rangle| \quad (3.65)$$
$$\geq |\langle BAx, y \rangle|,$$

for any $x, y \in H$, with $\|x\| = \|y\| = 1$.

Taking the supremum in (3.65) over $\|x\| = \|y\| = 1$ and noticing that

$$\sup_{\|x\|=1}|\langle Ax, x \rangle| = w(A)\,, \quad \sup_{\|x\|=\|y\|=1}|\langle Ax, y \rangle| = \|A\|\,, \quad \sup_{\|y\|=1}|\langle By, y \rangle| = w(B)\,,$$

$$\sup_{\|x\|=\|y\|=1}|\langle Bx, y \rangle| = \|B\|\,, \quad \sup_{\|x\|=\|y\|=1}|\langle x, y \rangle| = 1$$

and

$$\sup_{\|x\|=\|y\|=1}|\langle BAx, y \rangle| = \|BA\|\,,$$

we deduce the desired result (3.61).   ∎

Remark 124. It is an open problem whether or not the constant $\frac{1}{4}$ is best possible in the inequality (3.61).

A different approach is considered in the following result:

Theorem 125 (Dragomir [12], 2009). *With the assumptions from Theorem 123 we have the inequality*

$$\|BA\| \leq w(A)\,\|B\| + \frac{1}{4}|\beta - \alpha|\,(|\gamma + \delta| + |\gamma - \delta|)\,. \quad (3.66)$$

Proof. By the Schwarz inequality and taking into account the assumptions for the operators A and B we may state that

$$\left| \left\langle Ax - \langle Ax, x \rangle \, x, B^* y - \frac{\bar{\gamma} + \bar{\delta}}{2} y \right\rangle \right|$$

$$\leq \| Ax - \langle Ax, x \rangle \, x \| \left\| B^* y - \frac{\bar{\gamma} + \bar{\delta}}{2} y \right\|$$

$$\leq \left\| Ax - \frac{\alpha + \beta}{2} x \right\| \left\| B^* y - \frac{\bar{\gamma} + \bar{\delta}}{2} y \right\| \leq \frac{1}{4} |\beta - \alpha| \, |\gamma - \delta| \,, \qquad (3.67)$$

for any $x, y \in H$, with $\|x\| = \|y\| = 1$.

Now, since

$$\left\langle Ax - \langle Ax, x \rangle \, x, B^* y - \frac{\bar{\gamma} + \bar{\delta}}{2} y \right\rangle$$

$$= \langle BAx, y \rangle - \langle Ax, x \rangle \langle Bx, y \rangle - \frac{\gamma + \delta}{2} \langle Ax - \langle Ax, x \rangle \, x, y \rangle \,,$$

on taking the modulus in this equality we have

$$\left| \left\langle Ax - \langle Ax, x \rangle \, x, B^* y - \frac{\bar{\gamma} + \bar{\delta}}{2} y \right\rangle \right| \qquad (3.68)$$

$$\geq |\langle BAx, y \rangle| - |\langle Ax, x \rangle \langle Bx, y \rangle| - \left| \frac{\gamma + \delta}{2} \right| |\langle Ax - \langle Ax, x \rangle \, x, y \rangle| \,,$$

for any $x, y \in H$, with $\|x\| = \|y\| = 1$.

On making use of (3.67) and (3.68) we get

$$|\langle BAx, y \rangle| \qquad (3.69)$$

$$\leq |\langle Ax, x \rangle \langle Bx, y \rangle| + \left| \frac{\gamma + \delta}{2} \right| |\langle Ax - \langle Ax, x \rangle \, x, y \rangle| + \frac{1}{4} |\beta - \alpha| \, |\gamma - \delta|$$

$$\leq |\langle Ax, x \rangle \langle Bx, y \rangle| + \left| \frac{\gamma + \delta}{2} \right| \left\| Ax - \frac{\alpha + \beta}{2} x \right\| + \frac{1}{4} |\beta - \alpha| \, |\gamma - \delta|$$

$$\leq |\langle Ax, x \rangle \langle Bx, y \rangle| + \frac{1}{4} |\beta - \alpha| \, (|\gamma + \delta| + |\gamma - \delta|) \,,$$

for any $x, y \in H$, with $\|x\| = \|y\| = 1$.

Taking the supremum over $\|x\| = \|y\| = 1$ in (3.69) we deduce the desired inequality (3.66). ∎

In a similar manner we can state the following results as well:

Theorem 126 (Dragomir [12], 2009). *With the assumptions from Theorem 123 we have the inequality*

$$\|BA\| \le w(A)\|B\| + \frac{1}{2}|\gamma + \delta|(w(A) + \|A\|) + \frac{1}{4}|\beta - \alpha||\gamma - \delta|. \qquad (3.70)$$

Indeed, we observe that

$$\left\langle Ax - \langle Ax, x\rangle x, B^*y - \frac{\bar{\gamma} + \bar{\delta}}{2}y \right\rangle$$

$$= \langle BAx, y\rangle - \langle Ax, x\rangle \langle Bx, y\rangle - \frac{\gamma + \delta}{2}\langle Ax, y\rangle + \frac{\gamma + \delta}{2}\langle Ax, x\rangle \langle x, y\rangle$$

which produces the inequality

$$\left|\left\langle Ax - \langle Ax, x\rangle x, B^*y - \frac{\bar{\gamma} + \bar{\delta}}{2}y \right\rangle\right| + |\langle Ax, x\rangle \langle Bx, y\rangle|$$

$$+ \left|\frac{\gamma + \delta}{2}\right||\langle Ax, y\rangle| + \left|\frac{\gamma + \delta}{2}\right||\langle Ax, x\rangle||\langle x, y\rangle| \ge |\langle BAx, y\rangle|,$$

for any $x, y \in H$, with $\|x\| = \|y\| = 1$.

On utilizing the same argument as in the proof of the above theorem, we get the desired result (3.70). The details are omitted.

3.2.2 Related Results

The following result concerning an upper bound for the norm of the operator product may be stated.

Theorem 127 (Dragomir [12], 2009). *With the assumptions from Theorem 123 we have the inequality*

$$\|BA\| \le \frac{1}{4}|\beta - \alpha||\gamma - \delta| + \left\|\frac{\alpha + \beta}{2} \cdot B + \frac{\gamma + \delta}{2} \cdot A\right.$$

$$\left. - \frac{\alpha + \beta}{2} \cdot \frac{\gamma + \delta}{2} \cdot I\right\|$$

$$\le \frac{1}{4}|\beta - \alpha||\gamma - \delta| + \min\left\{\left|\frac{\alpha + \beta}{2}\right|(\|B\| + \frac{1}{2}|\beta - \alpha|),\right.$$

$$\left.\left|\frac{\gamma + \delta}{2}\right|(\|A\| + \frac{1}{2}|\gamma - \delta|)\right\}. \qquad (3.71)$$

Proof. By the Schwarz inequality and utilizing the assumptions about A and B we have

$$\left| \left\langle Ax - \frac{\alpha + \beta}{2}x, B^*y - \frac{\bar{\gamma} + \bar{\delta}}{2}y \right\rangle \right|$$

$$\leq \left\| Ax - \frac{\alpha + \beta}{2}x \right\| \left\| B^*y - \frac{\bar{\gamma} + \bar{\delta}}{2}y \right\| \leq \frac{1}{4}|\beta - \alpha| \, |\gamma - \delta|, \qquad (3.72)$$

for any $x, y \in H$, with $\|x\| = \|y\| = 1$.

Also, the following identity is of interest in itself

$$\left\langle Ax - \frac{\alpha + \beta}{2}x, B^*y - \frac{\bar{\gamma} + \bar{\delta}}{2}y \right\rangle \qquad (3.73)$$

$$= \langle BAx, y \rangle + \frac{\alpha + \beta}{2} \cdot \frac{\gamma + \delta}{2} \langle x, y \rangle - \frac{\alpha + \beta}{2} \langle Bx, y \rangle - \frac{\gamma + \delta}{2} \langle Ax, y \rangle,$$

for any $x, y \in H$, with $\|x\| = \|y\| = 1$.

This identity gives

$$\left\langle Ax - \frac{\alpha + \beta}{2}x, B^*y - \frac{\bar{\gamma} + \bar{\delta}}{2}y \right\rangle$$

$$+ \left\langle \frac{\alpha + \beta}{2} \cdot Bx + \frac{\gamma + \delta}{2} \cdot Ax - \frac{\alpha + \beta}{2} \cdot \frac{\gamma + \delta}{2}x, y \right\rangle = \langle BAx, y \rangle,$$

for any $x, y \in H$, with $\|x\| = \|y\| = 1$.

Taking the modulus and utilizing (3.72) we get

$$|\langle BAx, y \rangle| \leq \left| \left\langle Ax - \frac{\alpha + \beta}{2}x, B^*y - \frac{\bar{\gamma} + \bar{\delta}}{2}y \right\rangle \right|$$

$$+ \left| \left\langle \frac{\alpha + \beta}{2} \cdot Bx + \frac{\gamma + \delta}{2} \cdot Ax - \frac{\alpha + \beta}{2} \cdot \frac{\gamma + \delta}{2}x, y \right\rangle \right|$$

$$\leq \frac{1}{4}|\beta - \alpha| \, |\gamma - \delta|$$

$$+ \left\| \frac{\alpha + \beta}{2} \cdot Bx + \frac{\gamma + \delta}{2} \cdot Ax - \frac{\alpha + \beta}{2} \cdot \frac{\gamma + \delta}{2}x \right\|,$$

for any $x, y \in H$, with $\|x\| = \|y\| = 1$.

Finally, taking the supremum over $\|x\| = \|y\| = 1$, we deduce the first part of the desired inequality (3.71). The second part is obvious by the triangle inequality and by the assumptions on A and B. ∎

The following particular case also holds

Corollary 128. *Let $A \in B(H)$ and $\alpha, \beta \in \mathbb{K}$ be such that the transforms $C_{\alpha,\beta}(A)$ is accretive. Then*

$$\|A^2\| \le \frac{1}{4}|\beta - \alpha|^2 + \left|\frac{\alpha + \beta}{2}\right| \left\|2 \cdot A - \frac{\alpha + \beta}{2} \cdot I\right\|$$

$$\left(\le \frac{1}{4}|\beta - \alpha|^2 + \left|\frac{\alpha + \beta}{2}\right| \left(\|A\| + \frac{1}{2}|\beta - \alpha|\right)\right) \qquad (3.74)$$

and

$$\|A\|^2 \le \frac{1}{4}|\beta - \alpha|^2$$

$$+ \left\|\frac{\bar{\alpha} + \bar{\beta}}{2} \cdot A^* + \frac{\alpha + \beta}{2} \cdot A - \left|\frac{\alpha + \beta}{2}\right|^2 \cdot I\right\|$$

$$\left(\le \frac{1}{4}|\beta - \alpha|^2 + \left|\frac{\alpha + \beta}{2}\right| \left(\|A\| + \frac{1}{2}|\beta - \alpha|\right)\right), \qquad (3.75)$$

respectively.

The following result provides an approximation for the operator product in terms of some simpler quantities:

Theorem 129 (Dragomir [12], 2009). *With the assumptions from Theorem 123 we have the inequality*

$$\left\|BA - \frac{\alpha + \beta}{2} \cdot B - \frac{\gamma + \delta}{2} \cdot A + \frac{\alpha + \beta}{2} \cdot \frac{\gamma + \delta}{2} \cdot I\right\|$$

$$\le \frac{1}{4}|\beta - \alpha| |\gamma - \delta| . \qquad (3.76)$$

Proof. The identity (3.73) can be written in an equivalent form as

$$\left\langle Ax - \frac{\alpha + \beta}{2}x, B^*y - \frac{\bar{\gamma} + \bar{\delta}}{2}y \right\rangle$$

$$= \left\langle \left(BA - \frac{\alpha + \beta}{2} \cdot B - \frac{\gamma + \delta}{2} \cdot A + \frac{\alpha + \beta}{2} \cdot \frac{\gamma + \delta}{2} \cdot I\right)x, y \right\rangle, \qquad (3.77)$$

for any $x, y \in H$, with $\|x\| = \|y\| = 1$.

Taking the modulus and making use of the inequality (3.72) we get

$$\left|\left\langle \left(BA - \frac{\alpha + \beta}{2} \cdot B - \frac{\gamma + \delta}{2} \cdot A + \frac{\alpha + \beta}{2} \cdot \frac{\gamma + \delta}{2} \cdot I\right)x, y \right\rangle\right|$$

$$\le \frac{1}{4}|\beta - \alpha| |\gamma - \delta| ,$$

for any $x, y \in H$, with $\|x\| = \|y\| = 1$, which implies the desired result (3.76). ∎

Corollary 130. *Let $A \in B(H)$ and $\alpha, \beta \in \mathbb{K}$ be such that the transform $C_{\alpha,\beta}(A)$ is accretive, then*

$$\left\| A^2 - (\alpha + \beta) \cdot A + \left(\frac{\alpha + \beta}{2} \right)^2 \cdot I \right\| \leq \frac{1}{4} |\beta - \alpha|^2 \qquad (3.78)$$

and

$$\left\| A^* A - \frac{\alpha + \beta}{2} \cdot A^* - \frac{\bar{\alpha} + \bar{\beta}}{2} \cdot A + \left| \frac{\alpha + \beta}{2} \right|^2 \cdot I \right\| \leq \frac{1}{4} |\beta - \alpha|^2, \qquad (3.79)$$

respectively.

The following theorem provides an approximation for the operator

$$\frac{1}{2} \left(U^* U + U U^* \right)$$

when some information about the real or imaginary part of the operator U are given.

We recall that $U = \operatorname{Re}(U) + i \operatorname{Im}(U)$, i.e., $\operatorname{Re}(U) = \frac{1}{2}(U + U^*)$ and $\operatorname{Im}(U) = \frac{1}{2i}(U - U^*)$. For simplicity, we denote by A the real part of U and by B its imaginary part.

Theorem 131 (Dragomir [12], 2009). *Suppose that $a, b, c, d \in \mathbb{R}$ are such that $C_{a,c}(A)$ and $C_{b,d}(B)$ are accretive. Denote $\alpha := a + ib$ and $\beta := c + id \in \mathbb{C}$, then*

$$\left\| \frac{1}{2} \left(U^* U + U U^* \right) - \frac{\bar{\alpha} + \bar{\beta}}{2} \cdot U - \frac{\alpha + \beta}{2} \cdot U^* + \left| \frac{\alpha + \beta}{2} \right|^2 \cdot I \right\|$$

$$\leq \frac{1}{4} |\alpha - \beta|^2. \qquad (3.80)$$

Proof. It is well known that for any operator T with the Cartesian decomposition $T = C + iD$ we have

$$\frac{1}{2} \left(T^* T + T T^* \right) = C^2 + D^2. \qquad (3.81)$$

For any $z \in \mathbb{C}$ we also have the identity

$$\frac{1}{2} \left[(U - zI)(U^* - \bar{z}I) + (U^* - \bar{z}I)(U - zI) \right]$$

$$= \frac{1}{2} \left(U^* U + U U^* \right) - \bar{z} \cdot U - z \cdot U^* + |z|^2 \cdot I. \qquad (3.82)$$

For $z = \frac{\alpha + \beta}{2}$ we observe that

$$\operatorname{Re}(U - zI) = A - \frac{a + c}{2} \cdot I \qquad \text{and} \qquad \operatorname{Im}(U - zI) = B - \frac{b + d}{2} \cdot I$$

and utilizing the identities (3.81) and (3.82) we deduce

$$\left\| \frac{1}{2} \left(U^* U + U U^* \right) - \bar{z} \cdot U - z \cdot U^* + |z|^2 \cdot I \right\|$$

$$= \left\| \left(A - \frac{a + c}{2} \cdot I \right)^2 + \left(B - \frac{b + d}{2} \cdot I \right)^2 \right\|$$

$$\leq \left\| A - \frac{a + c}{2} \cdot I \right\|^2 + \left\| B - \frac{b + d}{2} \cdot I \right\|^2$$

$$\leq \frac{1}{4} \left[(c - a)^2 + (d - b)^2 \right] = \frac{1}{4} |\alpha - \beta|^2,$$

where for the last inequality we have used the fact that $C_{a,c} (A)$ and $C_{b,d} (B)$ are accretive. ∎

3.3 Power Inequalities for the Numerical Radius

3.3.1 Inequalities for a Product of Two Operators

Theorem 132 (Dragomir [16], 2009). *For any $A, B \in B (H)$ and $r \geq 1$, we have the inequality:*

$$w^r \left(B^* A \right) \leq \frac{1}{2} \left\| \left(A^* A \right)^r + \left(B^* B \right)^r \right\|. \tag{3.83}$$

The constant $\frac{1}{2}$ is best possible.

Proof. By the Schwarz inequality in the Hilbert space $(H; \langle \cdot, \cdot \rangle)$ we have

$$|\langle B^* A x, x \rangle| = |\langle A x, B x \rangle|$$

$$\leq \|A x\| \cdot \|B x\|$$

$$= \langle A^* A x, x \rangle^{\frac{1}{2}} \cdot \langle B^* B x, x \rangle^{\frac{1}{2}}, \qquad x \in H. \tag{3.84}$$

Utilizing the arithmetic mean-geometric mean inequality and then the convexity of the function $f (t) = t^r, r \geq 1$, we have successively

$$\langle A^* A x, x \rangle^{\frac{1}{2}} \cdot \langle B^* B x, x \rangle^{\frac{1}{2}} \leq \frac{\langle A^* A x, x \rangle + \langle B^* B x, x \rangle}{2} \tag{3.85}$$

$$\leq \left(\frac{\langle A^* A x, x \rangle^r + \langle B^* B x, x \rangle^r}{2} \right)^{\frac{1}{r}}$$

for any $x \in H$.

It is known that if P is a positive operator then for any $r \geq 1$ and $x \in H$ with $\|x\| = 1$ we have the inequality (see for instance [23])

$$\langle Px, x \rangle^r \leq \langle P^r x, x \rangle . \tag{3.86}$$

Applying this property to the positive operator $A^* A$ and $B^* B$, we deduce that

$$\left(\frac{\langle A^* Ax, x \rangle^r + \langle B^* Bx, x \rangle^r}{2} \right)^{\frac{1}{r}}$$

$$\leq \left(\frac{\langle (A^* A)^r x, x \rangle + \langle (B^* B)^r x, x \rangle}{2} \right)^{\frac{1}{r}}$$

$$= \left(\frac{\langle [(A^* A)^r + (B^* B)^r] x, x \rangle}{2} \right)^{\frac{1}{r}} \tag{3.87}$$

for any $x \in H$, $\|x\| = 1$.

Now, on making use of the inequalities (3.84), (3.85) and (3.87), we get the inequality:

$$\left| \langle (B^* A)^r x, x \rangle \right|^r \leq \frac{1}{2} \langle [(A^* A)^r + (B^* B)^r] x, x \rangle \tag{3.88}$$

for any $x \in H$, $\|x\| = 1$.

Taking the supremum over $x \in H$, $\|x\| = 1$ in (3.88) and since the operator $\left[(A^* A)^r + (B^* B)^r \right]$ is self-adjoint, we deduce the desired inequality (3.83).

For $r = 1$ and $B = A$, we get on both sides of (3.83) the same quantity $\|A\|^2$ which shows that the constant $\frac{1}{2}$ is best possible in general in the inequality (3.83). ∎

Corollary 133. *For any $A \in B(H)$ and $r \geq 1$ we have the inequalities:*

$$w^r(A) \leq \frac{1}{2} \left\| (A^* A)^r + I \right\| \tag{3.89}$$

and

$$w^r(A^2) \leq \frac{1}{2} \left\| (A^* A)^r + (AA^*)^r \right\| , \tag{3.90}$$

respectively.

A different approach is considered in the following result:

Theorem 134 (Dragomir [16], 2009). *For any $A, B \in B(H)$ and any $\alpha \in (0, 1)$ and $r \geq 1$, we have the inequality:*

$$w^{2r}(B^* A) \leq \left\| \alpha (A^* A)^{\frac{r}{\alpha}} + (1 - \alpha) (B^* B)^{\frac{r}{1-\alpha}} \right\| . \tag{3.91}$$

Proof. By Schwarz's inequality, we have

$$|\langle (B^*A)\, x, x\rangle|^2 \le \langle (A^*A)\, x, x\rangle \cdot \langle (B^*B)\, x, x\rangle \tag{3.92}$$

$$= \left\langle \left[(A^*A)^{\frac{1}{\alpha}}\right]^\alpha x, x\right\rangle \cdot \left\langle \left[(B^*B)^{\frac{1}{1-\alpha}}\right]^{1-\alpha} x, x\right\rangle,$$

for any $x \in H$.

It is well known that (see for instance [23]) if P is a positive operator and $q \in (0, 1]$ then for any $u \in H$, $\|u\| = 1$, we have

$$\langle P^q u, u\rangle \le \langle P u, u\rangle^q . \tag{3.93}$$

Applying this property to the positive operators $(A^*A)^{\frac{1}{\alpha}}$ and $(B^*B)^{\frac{1}{1-\alpha}}$ ($\alpha \in (0, 1)$), we have

$$\left\langle \left[(A^*A)^{\frac{1}{\alpha}}\right]^\alpha x, x\right\rangle \cdot \left\langle \left[(B^*B)^{\frac{1}{1-\alpha}}\right]^{1-\alpha} x, x\right\rangle$$

$$\le \left\langle (A^*A)^{\frac{1}{\alpha}} x, x\right\rangle^\alpha \cdot \left\langle (B^*B)^{\frac{1}{1-\alpha}} x, x\right\rangle^{1-\alpha} , \tag{3.94}$$

for any $x \in H$, $\|x\| = 1$.

Now, utilizing the weighted arithmetic mean-geometric mean inequality, i.e., $a^\alpha b^{1-\alpha} \le \alpha a + (1 - \alpha) b, \alpha \in (0, 1), a, b \ge 0$, we get

$$\left\langle (A^*A)^{\frac{1}{\alpha}} x, x\right\rangle^\alpha \cdot \left\langle (B^*B)^{\frac{1}{1-\alpha}} x, x\right\rangle^{1-\alpha}$$

$$\le \alpha \left\langle (A^*A)^{\frac{1}{\alpha}} x, x\right\rangle + (1 - \alpha) \left\langle (B^*B)^{\frac{1}{1-\alpha}} x, x\right\rangle \tag{3.95}$$

for any $x \in H$, $\|x\| = 1$.

Moreover, by the elementary inequality following from the convexity of the function $f(t) = t^r, r \ge 1$, namely

$$\alpha a + (1 - \alpha) b \le (\alpha a^r + (1 - \alpha) b^r)^{\frac{1}{r}} , \qquad \alpha \in (0, 1), \ a, b \ge 0,$$

we deduce that

$$\alpha \left\langle (A^*A)^{\frac{1}{\alpha}} x, x\right\rangle + (1 - \alpha) \left\langle (B^*B)^{\frac{1}{1-\alpha}} x, x\right\rangle \tag{3.96}$$

$$\le \left[\alpha \left\langle (A^*A)^{\frac{1}{\alpha}} x, x\right\rangle^r + (1 - \alpha) \left\langle (B^*B)^{\frac{1}{1-\alpha}} x, x\right\rangle^r\right]^{\frac{1}{r}}$$

$$\le \left[\alpha \left\langle (A^*A)^{\frac{r}{\alpha}} x, x\right\rangle + (1 - \alpha) \left\langle (B^*B)^{\frac{r}{1-\alpha}} x, x\right\rangle\right]^{\frac{1}{r}} ,$$

for any $x \in H$, $\|x\| = 1$, where, for the last inequality, we used the inequality (3.86) for the positive operators $(A^*A)^{\frac{1}{\alpha}}$ and $(B^*B)^{\frac{1}{1-\alpha}}$.

Now, on making use of the inequalities (3.92), (3.94), (3.95) and (3.96), we get

$$\left| \langle (B^*A)\, x, x \rangle \right|^{2r} \leq \left\langle \left[\alpha \left(A^*A \right)^{\frac{r}{\alpha}} + (1 - \alpha) \left(B^*B \right)^{\frac{r}{1-\alpha}} \right] x, x \right\rangle \tag{3.97}$$

for any $x \in H$, $\|x\| = 1$. Taking the supremum over $x \in H$, $\|x\| = 1$ in (3.97) produces the desired inequality (3.91). ∎

Remark 135. The particular case $\alpha = \frac{1}{2}$ produces the inequality

$$w^{2r} \left(B^*A \right) \leq \frac{1}{2} \left\| \left(A^*A \right)^{2r} + \left(B^*B \right)^{2r} \right\|, \tag{3.98}$$

for $r \geq 1$. Notice that $\frac{1}{2}$ is best possible in (3.98) since for $r = 1$ and $B = A$ we get on both sides of (3.98) the same quantity $\|A\|^4$.

Corollary 136. *For any $A \in B(H)$ and $\alpha \in (0, 1)$, $r \geq 1$, we have the inequalities*

$$w^{2r} (A) \leq \left\| \alpha \left(A^*A \right)^{\frac{r}{\alpha}} + (1 - \alpha) I \right\| \tag{3.99}$$

and

$$w^{2r} \left(A^2 \right) \leq \left\| \alpha \left(A^*A \right)^{\frac{r}{\alpha}} + (1 - \alpha) \left(AA^* \right)^{\frac{r}{1-\alpha}} \right\|, \tag{3.100}$$

respectively.
   Moreover, we have

$$\|A\|^{4r} \leq \left\| \alpha \left(A^*A \right)^{\frac{r}{\alpha}} + (1 - \alpha) \left(A^*A \right)^{\frac{r}{1-\alpha}} \right\|. \tag{3.101}$$

3.3.2 Inequalities for the Sum of Two Products

The following result may be stated:

Theorem 137 (Dragomir [16], 2009). *For any $A, B, C, D \in B(H)$ and $r, s \geq 1$ we have*

$$\left\| \frac{B^*A + D^*C}{2} \right\|^2 \leq \left\| \frac{(A^*A)^r + (C^*C)^r}{2} \right\|^{\frac{1}{r}}$$
$$\cdot \left\| \frac{(B^*B)^s + (D^*D)^s}{2} \right\|^{\frac{1}{s}}. \tag{3.102}$$

Proof. By the Schwarz inequality in the Hilbert space $(H; \langle \cdot, \cdot \rangle)$ we have

$$\left| \langle (B^*A + D^*C) x, y \rangle \right|^2$$
$$= \left| \langle B^*Ax, y \rangle + \langle D^*Cx, y \rangle \right|^2$$
$$\leq \left[\left| \langle B^*Ax, y \rangle \right| + \left| \langle D^*Cx, y \rangle \right| \right]^2$$
$$\leq \left[\langle A^*Ax, x \rangle^{\frac{1}{2}} \cdot \langle B^*By, y \rangle^{\frac{1}{2}} + \langle C^*Cx, x \rangle^{\frac{1}{2}} \cdot \langle D^*Dy, y \rangle^{\frac{1}{2}} \right]^2, \qquad (3.103)$$

for any $x, y \in H$.

Now, on utilizing the elementary inequality:

$$(ab + cd)^2 \leq \left(a^2 + c^2 \right) \left(b^2 + d^2 \right), \qquad a, b, c, d \in \mathbb{R},$$

we then conclude that:

$$\left[\langle A^*Ax, x \rangle^{\frac{1}{2}} \cdot \langle B^*By, y \rangle^{\frac{1}{2}} + \langle C^*Cx, x \rangle^{\frac{1}{2}} \cdot \langle D^*Dy, y \rangle^{\frac{1}{2}} \right]^2$$
$$\leq \left(\langle A^*Ax, x \rangle + \langle C^*Cx, x \rangle \right) \cdot \left(\langle B^*By, y \rangle + \langle D^*Dy, y \rangle \right), \qquad (3.104)$$

for any $x, y \in H$.

Now, on making use of a similar argument to the one in the proof of Theorem 132, we have for $r, s \geq 1$ that

$$\left(\langle A^*Ax, x \rangle + \langle C^*Cx, x \rangle \right) \cdot \left(\langle B^*By, y \rangle + \langle D^*Dy, y \rangle \right)$$
$$\leq 4 \left\langle \left[\frac{(A^*A)^r + (C^*C)^r}{2} \right] x, x \right\rangle^{\frac{1}{r}} \cdot \left\langle \left[\frac{(B^*B)^s + (D^*D)^s}{2} \right] y, y \right\rangle^{\frac{1}{s}} \qquad (3.105)$$

for any $x, y \in H$, $\|x\| = \|y\| = 1$.

Consequently, by (3.103)–(3.105), we have

$$\left| \left\langle \left[\frac{B^*A + D^*C}{2} \right] x, y \right\rangle \right|^2$$
$$\leq \left\langle \left[\frac{(A^*A)^r + (C^*C)^r}{2} \right] x, x \right\rangle^{\frac{1}{r}} \cdot \left\langle \left[\frac{(B^*B)^s + (D^*D)^s}{2} \right] y, y \right\rangle^{\frac{1}{s}} \qquad (3.106)$$

for any $x, y \in H$, $\|x\| = \|y\| = 1$.

Taking the supremum over $x, y \in H$, $\|x\| = \|y\| = 1$ we deduce the desired inequality (3.102). ∎

Remark 138. If $s = r$, then the inequality (3.102) is equivalent with:

$$\left\| \frac{B^*A + D^*C}{2} \right\|^{2r} \leq \left\| \frac{(A^*A)^r + (C^*C)^r}{2} \right\| \cdot \left\| \frac{(B^*B)^r + (D^*D)^r}{2} \right\|. \qquad (3.107)$$

Corollary 139. *For any $A, C \in B(H)$ we have*

$$\left\| \frac{A+C}{2} \right\|^{2r} \leq \left\| \frac{(A^*A)^r + (C^*C)^r}{2} \right\|, \qquad (3.108)$$

where $r \geq 1$. Also, we have

$$\left\| \frac{A^2 + C^2}{2} \right\|^{2} \leq \left\| \frac{(A^*A)^r + (C^*C)^r}{2} \right\|^{\frac{1}{r}} \cdot \left\| \frac{(AA^*)^s + (CC^*)^s}{2} \right\|^{\frac{1}{s}} \qquad (3.109)$$

for all $r, s \geq 1$ and in particular

$$\left\| \frac{A^2 + C^2}{2} \right\|^{2r} \leq \left\| \frac{(A^*A)^r + (C^*C)^r}{2} \right\| \cdot \left\| \frac{(AA^*)^r + (CC^*)^r}{2} \right\| \qquad (3.110)$$

for $r \geq 1$.

The inequality (3.108) follows from (3.102) for $B = D = I$, while the inequality (3.109) is obtained from the same inequality (3.102) for $B = A^*$ and $D = C^*$.

Another particular result of interest is the following one:

Corollary 140. *For any $A, B \in B(H)$ we have*

$$\left\| \frac{B^*A + A^*B}{2} \right\|^{2} \leq \left\| \frac{(A^*A)^r + (B^*B)^r}{2} \right\|^{\frac{1}{r}} \cdot \left\| \frac{(A^*A)^s + (B^*B)^s}{2} \right\|^{\frac{1}{s}} \qquad (3.111)$$

for $r, s \geq 1$ and, in particular,

$$\left\| \frac{B^*A + A^*B}{2} \right\|^{r} \leq \left\| \frac{(A^*A)^r + (B^*B)^r}{2} \right\| \qquad (3.112)$$

for any $r \geq 1$.

The inequality (3.110) follows from (3.102) for $D = A$ and $C = B$.

Another particular case that might be of interest is the following one.

Corollary 141. *For any $A, D \in B(H)$ we have*

$$\left\| \frac{A+D}{2} \right\|^{2} \leq \left\| \frac{(A^*A)^r + I}{2} \right\|^{\frac{1}{r}} \cdot \left\| \frac{(DD^*)^s + I}{2} \right\|^{\frac{1}{s}}, \qquad (3.113)$$

where $r, s \geq 1$. In particular

$$\|A\|^2 \leq \left\| \frac{(A^*A)^r + I}{2} \right\|^{\frac{1}{r}} \cdot \left\| \frac{(AA^*)^s + I}{2} \right\|^{\frac{1}{s}}. \qquad (3.114)$$

Moreover, for any $r \geq 1$, we have

$$\|A\|^{2r} \leq \left\| \frac{(A^*A)^r + I}{2} \right\| \cdot \left\| \frac{(AA^*)^r + I}{2} \right\|.$$

The proof is obvious by the inequality (3.102) on choosing $B = I$, $C = I$ and writing the inequality for D^* instead of D.

Remark 142. If $T \in B(H)$ and $T = A + iC$, i.e., A and C are its Cartesian decomposition, then we get from (3.108) that

$$\|T\|^{2r} \leq 2^{2r-1} \|A^{2r} + C^{2r}\|,$$

for any $r \geq 1$.

Also, since $A = \text{Re}(T) = \frac{T+T^*}{2}$ and $C = \text{Im}(T) = \frac{T-T^*}{2i}$, then from (3.108) we get the following inequalities as well:

$$\|\text{Re}(T)\|^{2r} \leq \left\| \frac{(T^*T)^r + (TT^*)^r}{2} \right\|$$

and

$$\|\text{Im}(T)\|^{2r} \leq \left\| \frac{(T^*T)^r + (TT^*)^r}{2} \right\|$$

for any $r \geq 1$.

In terms of the *Euclidean radius* of two operators $w_e(\cdot, \cdot)$, where, as in [8],

$$w_e(T, U) := \sup_{\|x\|=1} \left(|\langle Tx, x \rangle|^2 + |\langle Ux, x \rangle|^2 \right)^{\frac{1}{2}},$$

we have the following result as well.

Theorem 143 (Dragomir [16], 2009). *For any $A, B, C, D \in B(H)$ and $p, q > 1$ with $\frac{1}{p} + \frac{1}{q} = 1$, we have the inequality:*

$$w_e^2(B^*A, D^*C) \leq \|(A^*A)^p + (C^*C)^p\|^{1/p}$$
$$\cdot \|(B^*B)^q + (D^*D)^q\|^{1/q}. \tag{3.115}$$

Proof. For any $x \in H$, $\|x\| = 1$ we have the inequalities

$$|\langle B^*Ax, x \rangle|^2 + |\langle D^*Cx, x \rangle|^2$$
$$\leq \langle A^*Ax, x \rangle \cdot \langle B^*Bx, x \rangle + \langle C^*Cx, x \rangle \cdot \langle D^*Dx, x \rangle$$

$$\leq \left(\langle A^*Ax, x \rangle^p + \langle C^*Cx, x \rangle^p \right)^{1/p} \cdot \left(\langle B^*Bx, x \rangle^q + \langle D^*Dx, x \rangle^q \right)^{1/q}$$

$$\leq \left(\langle (A^*A)^p \, x, x \rangle + \langle (C^*C)^p \, x, x \rangle \right)^{1/p} \cdot \left(\langle (B^*B)^q \, x, x \rangle + \langle (D^*D)^q \, x, x \rangle \right)^{1/q}$$

$$\leq \langle \left[(A^*A)^p + (C^*C)^p \right] x, x \rangle^{1/p} \cdot \langle \left[(B^*B)^q + (D^*D)^q \right] x, x \rangle^{1/q} .$$

Taking the supremum over $x \in H$, $\|x\| = 1$ and noticing that the operators $(A^*A)^p + (C^*C)^p$ and $(B^*B)^q + (D^*D)^q$ are self-adjoint, we deduce the desired inequality (3.115). ∎

The following particular case is of interest.

Corollary 144. *For any $A, C \in B(H)$ and $p, q > 1$ with $\frac{1}{p} + \frac{1}{q} = 1$, we have*

$$w_e^2 (A, C) \leq 2^{\frac{1}{q}} \left\| (A^*A)^p + (C^*C)^p \right\|^{\frac{1}{p}} .$$

The proof follows from (3.115) for $B = D = I$.

Corollary 145. *For any $A, D \in B(H)$ and $p, q > 1$ with $\frac{1}{p} + \frac{1}{q} = 1$, we have*

$$w_e^2 (A, D) \leq \left\| (A^*A)^p + I \right\|^{\frac{1}{p}} \cdot \left\| (D^*D)^q + I \right\|^{\frac{1}{q}} .$$

3.3.3 Inequalities for the Commutator

The commutator of two bounded linear operators T and U is the operator $TU - UT$. For the usual norm $\|\cdot\|$ and for any two operators T and U, by using the triangle inequality and the submultiplicity of the norm, we can state the following inequality:

$$\|TU - UT\| \leq 2 \|U\| \|T\| . \tag{3.116}$$

In [15], the following result has been obtained as well

$$\|TU - UT\| \leq 2 \min \{ \|T\|, \|U\| \} \min \{ \|T - U\|, \|T + U\| \} . \tag{3.117}$$

By utilizing Theorem 137 we can state the following result for the numerical radius of the commutator.

Proposition 146 (Dragomir [16], 2009). *For any $T, U \in B(H)$ and $r, s \geq 1$ we have*

$$\|TU - UT\|^2 \leq 2^{2 - \frac{1}{r} - \frac{1}{s}} \left\| (T^*T)^r + (U^*U)^r \right\|^{\frac{1}{r}}$$

$$\cdot \left\| (TT^*)^s + (UU^*)^s \right\|^{\frac{1}{s}} . \tag{3.118}$$

Proof. Follows by Theorem 137 on choosing $B = T^*$, $A = U$, $D = -U^*$ and $C = T$. ∎

Remark 147. In particular, for $r = s$, we get from (3.118) that

$$\|TU - UT\|^{2r} \leq 2^{2r-2} \left\|\left(T^*T\right)^r + \left(U^*U\right)^r\right\| \cdot \left\|\left(TT^*\right)^r + \left(UU^*\right)^r\right\| \quad (3.119)$$

and for $r = 1$ we get

$$\|TU - UT\|^2 \leq \|T^*T + U^*U\| \cdot \|TT^* + UU^*\|. \quad (3.120)$$

For a bounded linear operator $T \in B(H)$, the self-commutator is the operator $T^*T - TT^*$. Observe that the operator $V := -i(T^*T - TT^*)$ is self-adjoint and $w(V) = \|V\|$, i.e.,

$$w\left(T^*T - TT^*\right) = \|T^*T - TT^*\|.$$

Now, utilizing (3.118) for $U = T^*$, we can state the following corollary.

Corollary 148. *For any $T \in B(H)$ we have the inequality:*

$$\|T^*T - TT^*\|^2 \leq 2^{2-\frac{1}{r}-\frac{1}{s}} \left\|\left(T^*T\right)^r + \left(TT^*\right)^r\right\|^{\frac{1}{r}}$$
$$\cdot \left\|\left(T^*T\right)^s + \left(TT^*\right)^s\right\|^{\frac{1}{s}}. \quad (3.121)$$

In particular, we have

$$\|T^*T - TT^*\|^r \leq 2^{r-1} \left\|\left(T^*T\right)^r + \left(TT^*\right)^r\right\|, \quad (3.122)$$

for any $r \geq 1$.

Moreover, for $r = 1$, we have

$$\|T^*T - TT^*\| \leq \|T^*T + TT^*\|. \quad (3.123)$$

3.4   A Functional Associated with Two Operators

3.4.1   Some Basic Facts

For two bounded linear operators A, B in the Hilbert space $(H, \langle \cdot, \cdot \rangle)$, we define the functional [17]

$$\mu(A, B) := \sup_{\|x\|=1} \{\|Ax\| \|Bx\|\} \, (\geq 0). \quad (3.124)$$

It is obvious that μ is symmetric and sub-additive in each variable, $\mu(A, A) = \|A\|^2$, $\mu(A, I) = \|A\|$, where I is the identity operator,

$$\mu(\alpha A, \beta B) = |\alpha \beta| \mu(A, B)$$

and

$$\mu(A, B) \le \|A\| \|B\|.$$

We also have the following inequalities

$$\mu(A, B) \ge w(B^* A) \tag{3.125}$$

and

$$\mu(A, B) \|A\| \|B\| \ge \mu(AB, BA). \tag{3.126}$$

The inequality (3.125) follows by the Schwarz inequality $\|Ax\| \|Bx\| \ge |\langle Ax, Bx \rangle|$, $x \in H$, while (3.126) can be obtained by multiplying the inequalities $\|ABx\| \le \|A\| \|Bx\|$ and $\|BAx\| \le \|B\| \|Ax\|$.

From (3.125) we also get

$$\|A\|^2 \ge \mu(A, A^*) \ge w(A^2) \tag{3.127}$$

for any A.

Motivated by the above results we establish in this section several inequalities for the functional $\mu(\cdot, \cdot)$ under various assumptions for the operators involved, including operators satisfying the uniform (α, β)−property and operators for which the transform $C_{\alpha, \beta}(\cdot, \cdot)$ is accretive.

3.4.2   General Inequalities

The following result concerning some general power operator inequalities may be stated:

Theorem 149 (Dragomir [17], 2010). *For any $A, B \in B(H)$ and $r \ge 1$ we have the inequality*

$$\mu^r(A, B) \le \frac{1}{2} \left\| (A^* A)^r + (B^* B)^r \right\|. \tag{3.128}$$

The constant $\frac{1}{2}$ is best possible.

Proof. Utilizing the arithmetic mean-geometric mean inequality and the convexity of the function $f(t) = t^r$ for $r \geq 1$ and $t \geq 0$ we have successively

$$\|Ax\|\|Bx\| \leq \frac{1}{2}\left[\langle A^*Ax, x\rangle + \langle B^*Bx, x\rangle\right]$$

$$\leq \left[\frac{\langle A^*Ax, x\rangle^r + \langle B^*Bx, x\rangle^r}{2}\right]^{\frac{1}{r}} \qquad (3.129)$$

for any $x \in H$.

It is well known that if P is a positive operator, then for any $r \geq 1$ and $x \in H$ with $\|x\| = 1$ we have the inequality (see for instance [22])

$$\langle Px, x\rangle^r \leq \langle P^r x, x\rangle. \qquad (3.130)$$

Applying this inequality to the positive operators A^*A and B^*B we deduce that

$$\left[\frac{\langle A^*Ax, x\rangle^r + \langle B^*Bx, x\rangle^r}{2}\right]^{\frac{1}{r}} \leq \left\langle \frac{\left[(A^*A)^r + (B^*B)^r\right]x}{2}, x\right\rangle^{\frac{1}{r}} \qquad (3.131)$$

for any $x \in H$ with $\|x\| = 1$.

Now, on making use of the inequalities (3.129) and (3.131), we get

$$\|Ax\|\|Bx\| \leq \left\langle \frac{\left[(A^*A)^r + (B^*B)^r\right]x}{2}, x\right\rangle^{\frac{1}{r}} \qquad (3.132)$$

for any $x \in H$ with $\|x\| = 1$. Taking the supremum over $x \in H$ with $\|x\| = 1$ we obtain the desired result (3.128).

For $r = 1$ and $B = A$ we get on both sides of (3.128) the same quantity $\|A\|^2$ which shows that the constant $\frac{1}{2}$ is best possible in general in the inequality (3.128). ∎

Corollary 150. *For any $A \in B(H)$ and $r \geq 1$ we have the inequalities*

$$\mu^r(A, A^*) \leq \frac{1}{2}\left\|(A^*A)^r + (AA^*)^r\right\| \qquad (3.133)$$

and

$$\|A\|^r \leq \frac{1}{2}\left\|(A^*A)^r + I\right\|, \qquad (3.134)$$

respectively.

The following similar result for powers of operators can be stated as well:

Theorem 151 (Dragomir [17], 2010). *For any $A, B \in B(H)$, any $\alpha \in (0, 1)$ and $r \geq 1$ we have the inequality*

$$\mu^{2r}(A, B) \leq \left\| \alpha \cdot (A^*A)^{\frac{r}{\alpha}} + (1-\alpha) \cdot (B^*B)^{\frac{r}{1-\alpha}} \right\|. \tag{3.135}$$

The inequality is sharp.

Proof. Observe that, for any $\alpha \in (0, 1)$, we have

$$\|Ax\|^2 \|Bx\|^2 = \langle (A^*A) x, x \rangle \langle (B^*B) x, x \rangle \tag{3.136}$$

$$= \left\langle \left[(A^*A)^{\frac{1}{\alpha}} \right]^{\alpha} x, x \right\rangle \left\langle \left[(B^*B)^{\frac{1}{1-\alpha}} \right]^{1-\alpha} x, x \right\rangle,$$

where $x \in H$.

It is well known that if P is a positive operator and $q \in (0, 1)$, then

$$\langle P^q x, x \rangle \leq \langle Px, x \rangle^q. \tag{3.137}$$

Applying this property to the positive operators $(A^*A)^{1/\alpha}$ and $(B^*B)^{1/(1-\alpha)}$, where $\alpha \in (0, 1)$, we have

$$\left\langle \left[(A^*A)^{\frac{1}{\alpha}} \right]^{\alpha} x, x \right\rangle \left\langle \left[(B^*B)^{\frac{1}{1-\alpha}} \right]^{1-\alpha} x, x \right\rangle \tag{3.138}$$

$$\leq \left\langle (A^*A)^{\frac{1}{\alpha}} x, x \right\rangle^{\alpha} \left\langle (B^*B)^{\frac{1}{1-\alpha}} x, x \right\rangle^{1-\alpha}$$

for any $x \in H$ with $\|x\| = 1$.

Now, on utilizing the weighted arithmetic mean-geometric mean inequality, i.e.

$$a^{\alpha} b^{1-\alpha} \leq \alpha a + (1-\alpha) b, \quad \text{where } \alpha \in (0, 1) \text{ and } a, b \geq 0,$$

we get

$$\left\langle (A^*A)^{\frac{1}{\alpha}} x, x \right\rangle^{\alpha} \left\langle (B^*B)^{\frac{1}{1-\alpha}} x, x \right\rangle^{1-\alpha}$$

$$\leq \alpha \cdot \left\langle (A^*A)^{\frac{1}{\alpha}} x, x \right\rangle + (1-\alpha) \cdot \left\langle (B^*B)^{\frac{1}{1-\alpha}} x, x \right\rangle \tag{3.139}$$

for any $x \in H$ with $\|x\| = 1$.

Moreover, by the elementary inequality

$$\alpha a + (1-\alpha) b \leq (\alpha a^r + (1-\alpha) b^r)^{\frac{1}{r}}, \quad \text{where } \alpha \in (0, 1) \text{ and } a, b \geq 0,$$

we have successively

$$\alpha \cdot \left\langle \left(A^* A \right)^{\frac{1}{\alpha}} x, x \right\rangle + (1 - \alpha) \cdot \left\langle \left(B^* B \right)^{\frac{1}{1-\alpha}} x, x \right\rangle$$

$$\leq \left[\alpha \cdot \left\langle \left(A^* A \right)^{\frac{1}{\alpha}} x, x \right\rangle^r + (1 - \alpha) \cdot \left\langle \left(B^* B \right)^{\frac{1}{1-\alpha}} x, x \right\rangle^r \right]^{\frac{1}{r}}$$

$$\leq \left[\alpha \cdot \left\langle \left(A^* A \right)^{\frac{r}{\alpha}} x, x \right\rangle + (1 - \alpha) \cdot \left\langle \left(B^* B \right)^{\frac{r}{1-\alpha}} x, x \right\rangle \right]^{\frac{1}{r}}, \qquad (3.140)$$

for any $x \in H$ with $\|x\| = 1$, where for the last inequality we have used the property (3.130) for the positive operators $(A^* A)^{1/\alpha}$ and $(B^* B)^{1/(1-\alpha)}$.

Now, on making use of the identity (3.136) and the inequalities (3.138)–(3.140) we get

$$\|Ax\|^2 \|Bx\|^2 \leq \left[\left\langle \left[\alpha \cdot \left(A^* A \right)^{\frac{r}{\alpha}} + (1 - \alpha) \cdot \left(B^* B \right)^{\frac{r}{1-\alpha}} \right] x, x \right\rangle \right]^{\frac{1}{r}}$$

for any $x \in H$ with $\|x\| = 1$. Taking the supremum over $x \in H$ with $\|x\| = 1$ we deduce the desired result (3.135).

Notice that the inequality is sharp since for $r = 1$ and $B = A$ we get on both sides of (3.135) the same quantity $\|A\|^4$. ■

Corollary 152. *For any $A \in B(H)$, any $\alpha \in (0, 1)$ and $r \geq 1$, we have the inequalities*

$$\mu^{2r} \left(A, A^* \right) \leq \left\| \alpha \cdot \left(A^* A \right)^{\frac{r}{\alpha}} + (1 - \alpha) \cdot \left(A A^* \right)^{\frac{r}{1-\alpha}} \right\|,$$

$$\|A\|^{2r} \leq \left\| \alpha \cdot \left(A^* A \right)^{\frac{r}{\alpha}} + (1 - \alpha) \cdot I \right\|$$

and

$$\|A\|^{4r} \leq \left\| \alpha \cdot \left(A^* A \right)^{\frac{r}{\alpha}} + (1 - \alpha) \cdot \left(A^* A \right)^{\frac{r}{1-\alpha}} \right\|,$$

respectively.

The following reverse of the inequality (3.125) may be stated as well:

Theorem 153 (Dragomir [17], 2010). *For any $A, B \in B(H)$ we have the inequalities*

$$(0 \leq) \mu (A, B) - w \left(B^* A \right) \leq \frac{1}{2} \|A - B\|^2 \qquad (3.141)$$

and

$$\mu \left(\frac{A + B}{2}, \frac{A - B}{2} \right) \leq \frac{1}{2} w \left(B^* A \right) + \frac{1}{4} \|A - B\|^2, \qquad (3.142)$$

respectively.

Proof. We have

$$\|Ax - Bx\|^2 = \|Ax\|^2 + \|Bx\|^2 - 2\operatorname{Re}\langle B^*Ax, x\rangle$$
$$\geq 2\|Ax\|\|Bx\| - 2|\langle B^*Ax, x\rangle|, \tag{3.143}$$

for any $x \in H, \|x\| = 1$, which gives the inequality

$$\|Ax\|\|Bx\| \leq |\langle B^*Ax, x\rangle| + \frac{1}{2}\|Ax - Bx\|^2,$$

for any $x \in H, \|x\| = 1$.

Taking the supremum over $\|x\| = 1$ we deduce the desired result (3.141).

By the parallelogram identity in the Hilbert space H we also have

$$\|Ax\|^2 + \|Bx\|^2 = \frac{1}{2}\left(\|Ax + Bx\|^2 + \|Ax - Bx\|^2\right)$$
$$\geq \|Ax + Bx\|\|Ax - Bx\|,$$

for any $x \in H$.

Combining this inequality with the first part of (3.143) we get

$$\|Ax + Bx\|\|Ax - Bx\| \leq \|Ax - Bx\|^2 + 2|\langle B^*Ax, x\rangle|,$$

for any $x \in H$. Taking the supremum in this inequality over $\|x\| = 1$ we deduce the desired result (3.142). ∎

Corollary 154. *Let $A \in B(H)$. If $\operatorname{Re}(A) := \frac{A+A^*}{2}$ and $\operatorname{Im}(A) := \frac{A-A^*}{2i}$ are the real and imaginary parts of A, then we have the inequalities*

$$(0 \leq)\, \mu\left(A, A^*\right) - w\left(A^2\right) \leq 2 \cdot \|\operatorname{Im}(A)\|^2$$

and

$$\mu\left(\operatorname{Re}(A), \operatorname{Im}(A)\right) \leq \frac{1}{2}w\left(A^2\right) + \|\operatorname{Im}(A)\|^2,$$

respectively.

Moreover, we have

$$(0 \leq)\, \mu\left(\operatorname{Re}(A), \operatorname{Im}(A)\right) - w\left(\operatorname{Re}(A)\operatorname{Im}(A)\right) \leq \frac{1}{2}\|A\|^2.$$

Corollary 155. *For any $A \in B(H)$ and $\lambda \in \mathbb{C}$ with $\lambda \neq 0$, we have the inequality (see also [10])*

$$(0 \leq)\, \|A\| - w(A) \leq \frac{1}{2|\lambda|}\|A - \lambda I\|^2. \tag{3.144}$$

For a bounded linear operator T consider the quantity $\ell(T) := \inf_{\|x\|=1}\|Tx\|$. We can state the following result as well.

Theorem 156 (Dragomir [17], 2010). *For any $A, B \in B(H)$ with $A \neq B$ and such that $\ell(B) \geq \|A - B\|$ we have*

$$(0 \leq) \mu^2(A, B) - w^2(B^*A) \leq \|A\|^2 \|A - B\|^2. \tag{3.145}$$

Proof. Denote $r := \|A - B\| > 0$. Then for any $x \in H$ with $\|x\| = 1$, we have $\|Bx\| \geq r$ and by the first part of (3.143) we can write that

$$\|Ax\|^2 + \left(\sqrt{\|Bx\|^2 - r^2}\right)^2 \leq 2|\langle B^*Ax, x\rangle| \tag{3.146}$$

for any $x \in H$ with $\|x\| = 1$.

On the other hand we have

$$\|Ax\|^2 + \left(\sqrt{\|Bx\|^2 - r^2}\right)^2 \geq 2 \cdot \|Ax\| \sqrt{\|Bx\|^2 - r^2} \tag{3.147}$$

for any $x \in H$ with $\|x\| = 1$.

Combining (3.146) with (3.147) we deduce

$$\|Ax\| \sqrt{\|Bx\|^2 - r^2} \leq |\langle B^*Ax, x\rangle|$$

which is clearly equivalent to

$$\|Ax\|^2 \|Bx\|^2 \leq |\langle B^*Ax, x\rangle|^2 + \|Ax\|^2 \|A - B\|^2 \tag{3.148}$$

for any $x \in H$ with $\|x\| = 1$. Taking the supremum in (3.148) over $x \in H$ with $\|x\| = 1$, we deduce the desired inequality (3.145). ∎

Corollary 157. *For any $A \in B(H)$ a non-self-adjoint operator in $B(H)$ and such that $\ell(A^*) \geq \|\text{Im}(A)\|$ we have*

$$(0 \leq) \mu^2(A, A^*) - w^2(A^2) \leq 4 \cdot \|A\|^2 \|\text{Im}(A)\|^2. \tag{3.149}$$

Corollary 158. *For any $A \in B(H)$ and $\lambda \in \mathbb{C}$ with $\lambda \neq 0$ and $|\lambda| \geq \|A - \lambda I\|$ we have the inequality (see also [10])*

$$(0 \leq) \|A\|^2 - w^2(A) \leq \frac{1}{|\lambda|^2} \cdot \|A\|^2 \|A - \lambda I\|^2$$

or, equivalently,

$$(0 \leq) \sqrt{1 - \frac{\|A - \lambda I\|^2}{|\lambda|^2}} \leq \frac{w(A)}{\|A\|} \, (\leq 1).$$

3.4.3 Operators Satisfying the Uniform (α, β)-Property

The following result that may be of interest in itself holds:

Lemma 159 (Dragomir [17], 2010). Let $T \in B(H)$ and $\alpha, \beta \in \mathbb{C}$ with $\alpha \neq \beta$. The following statements are equivalent:

(i) We have

$$\mathrm{Re} \langle \beta y - Tx, Tx - \alpha y \rangle \geq 0 \qquad (3.150)$$

for any $x, y \in H$ with $\|x\| = \|y\| = 1$;
(ii) We have

$$\left\| Tx - \frac{\alpha + \beta}{2} \cdot y \right\| \leq \frac{1}{2} |\alpha - \beta| \qquad (3.151)$$

for any $x, y \in H$ with $\|x\| = \|y\| = 1$.

Proof. This follows by the following identity

$$\mathrm{Re} \langle \beta y - Tx, Tx - \alpha y \rangle = \frac{1}{4} |\alpha - \beta|^2 - \left\| Tx - \frac{\alpha + \beta}{2} \cdot y \right\|^2,$$

that holds for any $x, y \in H$ with $\|x\| = \|y\| = 1$. ∎

Remark 160. For any operator $T \in B(H)$ if we choose $\alpha = a \|T\| (1 + 2i)$ and $\beta = a \|T\| (1 - 2i)$ with $a \geq 1$, then

$$\frac{\alpha + \beta}{2} = a \|T\| \quad \text{and} \quad \frac{|\alpha - \beta|}{2} = 2a \|T\|$$

showing that

$$\left\| Tx - \frac{\alpha + \beta}{2} \cdot y \right\| \leq \|Tx\| + \left| \frac{\alpha + \beta}{2} \right| \leq \|T\| + a \|T\|$$

$$\leq 2a \|T\| = \frac{1}{2} \cdot |\alpha - \beta|,$$

that holds for any $x, y \in H$ with $\|x\| = \|y\| = 1$, i.e., T satisfies the condition (3.150) with the scalars α and β given above.

Definition 161. For given $\alpha, \beta \in \mathbb{C}$ with $\alpha \neq \beta$ and $y \in H$ with $\|y\| = 1$, we say that the operator $T \in B(H)$ has the (α, β, y)-property if either (3.150) or, equivalently, (3.151) holds true for any $x \in H$ with $\|x\| = 1$. Moreover, if T has the (α, β, y)-property for any $y \in H$ with $\|y\| = 1$, then we say that this operator has the uniform (α, β)-property.

The following results may be stated:

Theorem 162 (Dragomir [17], 2010). *Let $A, B \in B(H)$ and $\alpha, \beta, \gamma, \delta \in \mathbb{K}$ with $\alpha \neq \beta$ and $\gamma \neq \delta$. For $y \in H$ with $\|y\| = 1$ assume that A^* has the (α, β, y)-property while B^* has the (γ, δ, y)-property. Then*

$$\left| \|Ay\| \|By\| - \|BA^*\| \right| \leq \frac{1}{4} |\beta - \alpha| |\gamma - \delta| . \tag{3.152}$$

Moreover, if A^* has the uniform (α, β)-property and B^* has the uniform (γ, δ)-property, then

$$|\mu (A, B) - \|BA^*\|| \leq \frac{1}{4} |\beta - \alpha| |\gamma - \delta| . \tag{3.153}$$

Proof. A^* has the (α, β, y)-property while B^* has the (γ, δ, y)-property; then on making use of Lemma 159 we have that

$$\left\| A^*x - \frac{\alpha + \beta}{2} \cdot y \right\| \leq \frac{1}{2} |\beta - \alpha|$$

and

$$\left\| B^*z - \frac{\gamma + \delta}{2} \cdot y \right\| \leq \frac{1}{2} |\gamma - \delta|$$

for any $x, z \in H$ with $\|x\| = \|z\| = 1$.

Now, we make use of the following Grüss type inequality for vectors in inner product spaces obtained by the author in [1] (see also [2] or [6, p. 43]):

Let $(H, \langle \cdot, \cdot \rangle)$ be an inner product space over the real or complex number field $\mathbb{K}$, $u, v, e \in H$, $\|e\| = 1$, and $\alpha, \beta, \gamma, \delta \in \mathbb{K}$ such that

$$\text{Re} \langle \beta e - u, u - \alpha e \rangle \geq 0, \qquad \text{Re} \langle \delta e - v, v - \gamma e \rangle \geq 0 \tag{3.154}$$

or, equivalently,

$$\left\| u - \frac{\alpha + \beta}{2} e \right\| \leq \frac{1}{2} |\beta - \alpha| , \qquad \left\| v - \frac{\gamma + \delta}{2} e \right\| \leq \frac{1}{2} |\delta - \gamma| . \tag{3.155}$$

Then

$$|\langle u, v \rangle - \langle u, e \rangle \langle e, v \rangle| \leq \frac{1}{4} |\beta - \alpha| |\delta - \gamma| . \tag{3.156}$$

Applying (3.156) for $u = A^*x$, $v = B^*z$ and $e = y$ we deduce

$$|\langle BA^*x, z \rangle - \langle x, Ay \rangle \langle By, z \rangle| \leq \frac{1}{4} |\beta - \alpha| |\delta - \gamma| , \tag{3.157}$$

for any $x, z \in H$, $\|x\| = \|z\| = 1$, which is an inequality of interest in itself.

Observing that

$$\left| |\langle BA^*x, z \rangle| - |\langle x, Ay \rangle \langle z, By \rangle| \right| \leq |\langle BA^*x, z \rangle - \langle x, Ay \rangle \langle By, z \rangle|,$$

then by (3.157) we deduce the inequality

$$\left| |\langle BA^*x, z \rangle| - |\langle x, Ay \rangle \langle z, By \rangle| \right| \leq \frac{1}{4} |\beta - \alpha| |\delta - \gamma|$$

for any $x, z \in H$, $\|x\| = \|z\| = 1$. This is equivalent to the following two inequalities

$$|\langle BA^*x, z \rangle| \leq |\langle x, Ay \rangle \langle z, By \rangle| + \frac{1}{4} |\beta - \alpha| |\delta - \gamma| \qquad (3.158)$$

and

$$|\langle x, Ay \rangle \langle z, By \rangle| \leq |\langle BA^*x, z \rangle| + \frac{1}{4} |\beta - \alpha| |\delta - \gamma| \qquad (3.159)$$

for any $x, z \in H$, $\|x\| = \|z\| = 1$.

Taking the supremum over $x, z \in H$, $\|x\| = \|z\| = 1$, in (3.158) and (3.159) we get the inequalities

$$\|BA^*\| \leq \|Ay\| \|By\| + \frac{1}{4} |\beta - \alpha| |\delta - \gamma| \qquad (3.160)$$

and

$$\|Ay\| \|By\| \leq \|BA^*\| + \frac{1}{4} |\beta - \alpha| |\delta - \gamma|, \qquad (3.161)$$

which are clearly equivalent to (3.152).

Now, if A^* has the uniform (α, β)-property and B^* has the uniform (γ, δ)-property, then the inequalities (3.160) and (3.161) hold for any $y \in H$ with $\|y\| = 1$. Taking the supremum over $y \in H$ with $\|y\| = 1$ in these inequalities we deduce

$$\|BA^*\| \leq \mu(A, B) + \frac{1}{4} |\beta - \alpha| |\delta - \gamma|$$

and

$$\mu(A, B) \leq \|BA^*\| + \frac{1}{4} |\beta - \alpha| |\delta - \gamma|,$$

which are equivalent to (3.153). ∎

Corollary 163. *Let $A \in B(H)$ and $\alpha, \beta, \gamma, \delta \in \mathbb{K}$ with $\alpha \neq \beta$ and $\gamma \neq \delta$. For $y \in H$ with $\|y\| = 1$ assume that A has the (α, β, y)-property while A^* has the (γ, δ, y)-property. Then*

$$\left| \|A^* y\| \, \|Ay\| - \|A^2\| \right| \leq \frac{1}{4} |\beta - \alpha| \, |\gamma - \delta| \,.$$

Moreover, if A has the uniform (α, β)-property and A^* has the uniform (γ, δ)-property, then

$$\left| \mu\left(A, A^*\right) - \|A^2\| \right| \leq \frac{1}{4} |\beta - \alpha| \, |\gamma - \delta| \,.$$

The following results may be stated as well:

Theorem 164 (Dragomir [17], 2010). *Let $A, B \in B(H)$ and $\alpha, \beta, \gamma, \delta \in \mathbb{K}$ with $\alpha + \beta \neq 0$ and $\gamma + \delta \neq 0$. For $y \in H$ with $\|y\| = 1$ assume that A^* has the (α, β, y)-property while B^* has the (γ, δ, y)-property. Then*

$$\left| \|Ay\| \, \|By\| - \|BA^*\| \right|$$

$$\leq \frac{1}{4} \cdot \frac{|\beta - \alpha| \, |\delta - \gamma|}{\sqrt{|\beta + \alpha|} \, |\delta + \gamma|} \sqrt{(\|A\| + \|Ay\|)(\|B\| + \|By\|)}. \tag{3.162}$$

Moreover, if A^* has the uniform (α, β)-property and B^* has the uniform (γ, δ)-property, then

$$\left| \mu\left(A, B\right) - \|BA^*\| \right| \leq \frac{1}{2} \cdot \frac{|\beta - \alpha| \, |\delta - \gamma|}{\sqrt{|\beta + \alpha|} \, |\delta + \gamma|} \sqrt{\|A\| \, \|B\|}. \tag{3.163}$$

Proof. We make use of the following inequality obtained by the author in [3] (see also [6, p. 65]):

Let $(H, \langle \cdot, \cdot \rangle)$ be an inner product space over the real or complex number field $\mathbb{K}$, $u, v, e \in H$, $\|e\| = 1$, and $\alpha, \beta, \gamma, \delta \in \mathbb{K}$ with $\alpha + \beta \neq 0$ and $\gamma + \delta \neq 0$ such that

$$\text{Re} \langle \beta e - u, u - \alpha e \rangle \geq 0, \qquad \text{Re} \langle \delta e - v, v - \gamma e \rangle \geq 0$$

or, equivalently,

$$\left\| u - \frac{\alpha + \beta}{2} e \right\| \leq \frac{1}{2} |\beta - \alpha| \,, \qquad \left\| v - \frac{\gamma + \delta}{2} e \right\| \leq \frac{1}{2} |\delta - \gamma| \,.$$

Then

$$| \langle u, v \rangle - \langle u, e \rangle \, \langle e, v \rangle |$$

$$\leq \frac{1}{4} \cdot \frac{|\beta - \alpha| \, |\delta - \gamma|}{\sqrt{|\beta + \alpha|} \, |\delta + \gamma|} \sqrt{(\|u\| + |\langle u, e \rangle|)(\|v\| + |\langle v, e \rangle|)}. \tag{3.164}$$

Applying (3.164) for $u = A^*x$, $v = B^*z$ and $e = y$ we deduce

$$|\langle BA^*x, z \rangle - \langle x, Ay \rangle \langle By, z \rangle|$$

$$\leq \frac{1}{4} \cdot \frac{|\beta - \alpha| |\delta - \gamma|}{\sqrt{|\beta + \alpha| |\delta + \gamma|}} \sqrt{(\|A^*x\| + |\langle x, Ay \rangle|)(\|B^*z\| + |\langle z, By \rangle|)}$$

for any $x, y, z \in H$, $\|x\| = \|y\| = \|z\| = 1$.

Now, on making use of a similar argument to the one from the proof of Theorem 162, we deduce the desired results (3.162) and (3.163). The details are omitted.   ∎

Corollary 165. *Let $A \in B(H)$ and $\alpha, \beta, \gamma, \delta \in \mathbb{K}$ with $\alpha + \beta \neq 0$ and $\gamma + \delta \neq 0$. For $y \in H$ with $\|y\| = 1$ assume that A has the (α, β, y)-property while A^* has the (γ, δ, y)-property. Then*

$$\left| \|A^*y\| \|Ay\| - \|A^2\| \right|$$

$$\leq \frac{1}{4} \cdot \frac{|\beta - \alpha| |\delta - \gamma|}{\sqrt{|\beta + \alpha| |\delta + \gamma|}} \sqrt{(\|A\| + \|A^*y\|)(\|A\| + \|Ay\|)}.$$

Moreover, if A has the uniform (α, β)-property and A^* has the uniform (γ, δ)-property, then

$$\left| \mu(A, A^*) - \|A^2\| \right| \leq \frac{1}{2} \cdot \frac{|\beta - \alpha| |\delta - \gamma|}{\sqrt{|\beta + \alpha| |\delta + \gamma|}} \|A\|.$$

3.4.4   The Transform $C_{\alpha, \beta}(\cdot, \cdot)$ and Other Inequalities

For two given operators $T, U \in B(H)$ and two given scalars $\alpha, \beta \in \mathbb{C}$ consider the transform

$$C_{\alpha, \beta}(T, U) = (T^* - \bar{\alpha}U^*)(\beta U - T).$$

This transform generalizes the transform $C_{\alpha, \beta}(T) := (T^* - \bar{\alpha}I)(\beta I - T) = C_{\alpha, \beta}(T, I)$, where I is the identity operator, which has been introduced before in order to provide some generalizations of the well-known Kantorovich inequality for operators in Hilbert spaces.

We recall that a bounded linear operator T on the complex Hilbert space $(H, \langle \cdot, \cdot \rangle)$ is called *accretive* if $\text{Re}\langle Ty, y \rangle \geq 0$ for any $y \in H$.

Utilizing the following identity

$$\operatorname{Re}\langle C_{\alpha,\beta}(T,U)x,x\rangle = \operatorname{Re}\langle C_{\beta,\alpha}(T,U)x,x\rangle$$

$$= \frac{1}{4}|\beta-\alpha|^2\|Ux\|^2 - \left\|Tx - \frac{\alpha+\beta}{2}\cdot Ux\right\|^2, \quad (3.165)$$

that holds for any scalars α, β and any vector $x \in H$, we can give a simple characterization result that is useful in the following:

Lemma 166 (Dragomir [17], 2010). *For $\alpha, \beta \in \mathbb{C}$ and $T, U \in B(H)$ the following statements are equivalent:*

(i) the transform $C_{\alpha,\beta}(T,U)$ $\big(or, equivalently, C_{\beta,\alpha}(T,U)\big)$ is accretive;
(ii) we have the norm inequality

$$\left\|Tx - \frac{\alpha+\beta}{2}\cdot Ux\right\| \le \frac{1}{2}|\beta-\alpha|\,\|Ux\| \quad (3.166)$$

for any $x \in H$.

As a consequence of the above lemma we can state

Corollary 167. *Let $\alpha, \beta \in \mathbb{C}$ and $T, U \in B(H)$. If $C_{\alpha,\beta}(T,U)$ is accretive, then*

$$\left\|T - \frac{\alpha+\beta}{2}\cdot U\right\| \le \frac{1}{2}|\beta-\alpha|\,\|U\|. \quad (3.167)$$

Remark 168. In order to give examples of linear operators $T, U \in B(H)$ and numbers $\alpha, \beta \in \mathbb{C}$ such that the transform $C_{\alpha,\beta}(T,U)$ is accretive, it suffices to select two bounded linear operators S and V and the complex numbers z, w $(w \ne 0)$ with the property that $\|Sx - zVx\| \le |w|\,\|Vx\|$ for any $x \in H$, and by choosing $T = S, U = V, \alpha = \frac{1}{2}(z+w)$ and $\beta = \frac{1}{2}(z-w)$ we observe that T and U satisfy (3.166), i.e., $C_{\alpha,\beta}(T,U)$ is accretive.

We are able now to give the following result concerning other reverse inequalities for the case when the involved operators satisfy the accretivity property described above.

Theorem 169 (Dragomir [17], 2010). *Let $\alpha, \beta \in \mathbb{C}$ and $A, B \in B(H)$. If $C_{\alpha,\beta}(A,B)$ is accretive, then*

$$(0 \le)\,\mu^2(A,B) - w^2(B^*A) \le \frac{1}{4}\cdot|\beta-\alpha|^2\,\|B\|^4. \quad (3.168)$$

Moreover, if $\alpha + \beta \ne 0$, then

$$(0 \le)\,\mu(A,B) - w(B^*A) \le \frac{1}{4}\cdot\frac{|\beta-\alpha|^2}{|\beta+\alpha|}\,\|B\|^2. \quad (3.169)$$

In addition, if $\mathrm{Re}\left(\alpha\bar{\beta}\right) > 0$ *and* $B^*A \neq 0$, *then also*

$$(1 \leq) \frac{\mu(A, B)}{w(B^*A)} \leq \frac{1}{2} \cdot \frac{|\beta + \alpha|}{\sqrt{\mathrm{Re}\left(\alpha\bar{\beta}\right)}} \tag{3.170}$$

and

$$(0 \leq) \mu^2(A, B) - w^2\left(B^*A\right)$$
$$\leq \left(|\beta + \alpha| - 2 \cdot \sqrt{\mathrm{Re}\left(\alpha\bar{\beta}\right)}\right) w\left(B^*A\right)\|B\|^2, \tag{3.171}$$

respectively.

Proof. By Lemma 166, since $C_{\alpha,\beta}(A, B)$ is accretive, then

$$\left\|Ax - \frac{\alpha + \beta}{2} \cdot Bx\right\| \leq \frac{1}{2}|\beta - \alpha|\,\|Bx\| \tag{3.172}$$

for any $x \in H$.

We utilize the following reverse of the Schwarz inequality in inner product spaces obtained by the author in [5] (see also [6, p. 4]):

If $\gamma, \Gamma \in \mathbb{K}$ ($\mathbb{K} = \mathbb{C}, \mathbb{R}$) and $u, v \in H$ are such that

$$\mathrm{Re}\,\langle \Gamma v - u, u - \gamma v\rangle \geq 0 \tag{3.173}$$

or, equivalently,

$$\left\|u - \frac{\gamma + \Gamma}{2} \cdot v\right\| \leq \frac{1}{2}|\Gamma - \gamma|\,\|v\|, \tag{3.174}$$

then

$$0 \leq \|u\|^2\,\|v\|^2 - |\langle u, v\rangle|^2 \leq \frac{1}{4}|\Gamma - \gamma|^2\,\|v\|^4. \tag{3.175}$$

Now, on making use of (3.175) for $u = Ax$, $v = Bx$, $x \in H$, $\|x\| = 1$ and $\gamma = \alpha$, $\Gamma = \beta$, we can write the inequality

$$\|Ax\|^2\,\|Bx\|^2 \leq |\langle B^*Ax, x\rangle|^2 + \frac{1}{4}|\beta - \alpha|^2\,\|Bx\|^4,$$

for any $x \in H$, $\|x\| = 1$. Taking the supremum over $\|x\| = 1$ in this inequality produces the desired result (3.168).

Now, by utilizing the result from [3] (see also [6, p. 29]), namely:

If $\gamma, \Gamma \in \mathbb{K}$ with $\gamma + \Gamma \neq 0$ and $u, v \in H$ are such that either (3.173) or, equivalently, (3.174) holds true, then

$$0 \leq \|u\| \, \|v\| - |\langle u, v \rangle| \leq \frac{1}{4} \cdot \frac{|\Gamma - \gamma|^2}{|\Gamma + \gamma|} \|v\|^2. \tag{3.176}$$

Now, on making use of (3.176) for $u = Ax$, $v = Bx$, $x \in H$, $\|x\| = 1$ and $\gamma = \alpha$, $\Gamma = \beta$ and using the same procedure outlined above, we deduce the second inequality (3.169).

The inequality (3.170) follows from the result presented below obtained in [4] (see also [6, p. 21]):

If $\gamma, \Gamma \in \mathbb{K}$ with $\operatorname{Re}(\Gamma \bar{\gamma}) > 0$ and $u, v \in H$ are such that either (3.173) or, equivalently, (3.174) holds true, then

$$\|u\| \, \|v\| \leq \frac{1}{2} \cdot \frac{|\Gamma + \gamma|}{\sqrt{\operatorname{Re}(\Gamma \bar{\gamma})}} |\langle u, v \rangle|, \tag{3.177}$$

by choosing $u = Ax$, $v = Bx$, $x \in H$, $\|x\| = 1$ and $\gamma = \alpha$, $\Gamma = \beta$ and taking the supremum over $\|x\| = 1$.

Finally, on making use of the inequality (see [10])

$$\|u\|^2 \, \|v\|^2 - |\langle u, v \rangle|^2 \leq \left(|\Gamma + \gamma| - 2\sqrt{\operatorname{Re}(\Gamma \bar{\gamma})} \right) |\langle u, v \rangle| \, \|v\|^2 \tag{3.178}$$

that is valid provided $\gamma, \Gamma \in \mathbb{K}$ with $\operatorname{Re}(\Gamma \bar{\gamma}) > 0$ and $u, v \in H$ are such that either (3.173) or, equivalently, (3.174) holds true, we obtain the last inequality (3.171). The details are omitted. ∎

Remark 170. Let $M, m > 0$ and $A, B \in B(H)$. If $C_{m,M}(A, B)$ is accretive, then

$$(0 \leq) \mu^2(A, B) - w^2(B^*A) \leq \frac{1}{4} \cdot (M - m)^2 \|B\|^4,$$

$$(0 \leq) \mu(A, B) - w(B^*A) \leq \frac{1}{4} \cdot \frac{(M - m)^2}{m + M} \|B\|^2,$$

$$(1 \leq) \frac{\mu(A, B)}{w(B^*A)} \leq \frac{1}{2} \cdot \frac{m + M}{\sqrt{mM}}$$

and

$$(0 \leq) \mu^2(A, B) - w^2(B^*A) \leq \left(\sqrt{M} - \sqrt{m} \right)^2 w(B^*A) \|B\|^2,$$

respectively.

Corollary 171. *Let $\alpha, \beta \in \mathbb{C}$ and $A \in B(H)$. If $C_{\alpha,\beta}(A, A^*)$ is accretive, then*

$$(0 \leq) \mu^2(A, A^*) - w^2(A^2) \leq \frac{1}{4} \cdot |\beta - \alpha|^2 \|A\|^4.$$

Moreover, if $\alpha + \beta \neq 0$, *then*

$$(0 \leq) \mu \left(A, A^* \right) - w \left(A^2 \right) \leq \frac{1}{4} \cdot \frac{|\beta - \alpha|^2}{|\beta + \alpha|} \, \|A\|^2 \, .$$

In addition, if $\operatorname{Re} \left(\alpha \bar{\beta} \right) > 0$ *and* $A^2 \neq 0$, *then also*

$$(1 \leq) \frac{\mu \left(A, A^* \right)}{w \left(A^2 \right)} \leq \frac{1}{2} \cdot \frac{|\beta + \alpha|}{\sqrt{\operatorname{Re} \left(\alpha \bar{\beta} \right)}}$$

and

$$(0 \leq) \mu^2 \left(A, A^* \right) - w^2 \left(A^2 \right) \leq \left(|\beta + \alpha| - 2 \cdot \sqrt{\operatorname{Re} \left(\alpha \bar{\beta} \right)} \right) w \left(A^2 \right) \|A\|^2 \, ,$$

respectively.

Remark 172. In a similar manner, if $N, n > 0$, $A \in B(H)$ and $C_{n,N} \left(A, A^* \right)$ is accretive, then

$$(0 \leq) \mu^2 \left(A, A^* \right) - w^2 \left(A^2 \right) \leq \frac{1}{4} \cdot (N - n)^2 \|A\|^4 \, ,$$

$$(0 \leq) \mu \left(A, A^* \right) - w \left(A^2 \right) \leq \frac{1}{4} \cdot \frac{(N - n)^2}{n + N} \|A\|^2 \, ,$$

$$(1 \leq) \frac{\mu \left(A, A^* \right)}{w \left(A^2 \right)} \leq \frac{1}{2} \cdot \frac{n + N}{\sqrt{nN}} \qquad \text{(for } A^2 \neq 0)$$

and

$$(0 \leq) \mu^2 \left(A, A^* \right) - w^2 \left(A^2 \right) \leq \left(\sqrt{N} - \sqrt{n} \right)^2 w \left(A^2 \right) \|A\|^2 \, ,$$

respectively.

3.5   Some Inequalities of the Grüss Type

3.5.1   *Additive and Multiplicative Grüss Type Inequalities*

Motivated by the natural questions that arise, in order to compare the quantity $w \left(AB \right)$ with other expressions comprising the norm or the numerical radius of the involved operators A and B (or certain expressions constructed with these operators), we establish in this section some natural inequalities of the form

$$w\left(BA\right) \le w\left(A\right)w\left(B\right) + K_1 \quad \text{(additive Grüss type inequality)}$$

or

$$\frac{w\left(BA\right)}{w\left(A\right)w\left(B\right)} \le K_2 \quad \text{(multiplicative Grüss type inequality)}$$

where K_1 and K_2 are specified and desirably simple constants (depending on the given operators A and B).

Applications in providing upper bounds for the nonnegative quantities

$$\|A\|^2 - w^2\left(A\right) \quad \text{and} \quad w^2\left(A\right) - w(A^2)$$

and the *super unitary* quantities

$$\frac{\|A\|^2}{w^2\left(A\right)} \quad \text{and} \quad \frac{w^2\left(A\right)}{w(A^2)}$$

are also given.

3.5.2   Numerical Radius Inequalities of Grüss Type

For the complex numbers α, β and the bounded linear operator T, we define the following transform:

$$C_{\alpha,\beta}\left(T\right) := \left(T^* - \overline{\alpha}I\right)\left(\beta I - T\right), \tag{3.179}$$

where by T^* we denote the adjoint of T.

The following results compare the quantities $w\left(AB\right)$ and $w\left(A\right)w\left(B\right)$ provided that some information about the transforms $C_{\alpha,\beta}\left(A\right)$ and $C_{\gamma,\delta}\left(B\right)$ are available, where $\alpha, \beta, \gamma, \delta \in \mathbb{K}$.

Theorem 173 (Dragomir [14], 2008). *Let $A, B \in B(H)$ and $\alpha, \beta, \gamma, \delta \in \mathbb{K}$ be such that the transforms $C_{\alpha,\beta}\left(A\right)$ and $C_{\gamma,\delta}\left(B\right)$ are accretive, then*

$$w\left(BA\right) \le w\left(A\right)w\left(B\right) + \frac{1}{4}\left|\beta - \alpha\right|\left|\gamma - \delta\right|. \tag{3.180}$$

Proof. Since $C_{\alpha,\beta}\left(A\right)$ and $C_{\gamma,\delta}\left(B\right)$ are accretive, then we have that

$$\left\|Ax - \frac{\alpha + \beta}{2}x\right\| \le \frac{1}{2}\left|\beta - \alpha\right|$$

and

$$\left\|B^*x - \frac{\overline{\gamma} + \overline{\delta}}{2}x\right\| \le \frac{1}{2}\left|\overline{\gamma} - \overline{\delta}\right|$$

for any $x \in H$, $\|x\| = 1$.

Now, we make use of the following Grüss type inequality for vectors in inner product spaces obtained by the author in [1] (see also [2] or [6, p. 43]):

Let $(H, \langle \cdot, \cdot \rangle)$ be an inner product space over the real or complex number field $\mathbb{K}$, $u, v, e \in H$, $\|e\| = 1$, and $\alpha, \beta, \gamma, \delta \in \mathbb{K}$ such that

$$\text{Re} \langle \beta e - u, u - \alpha e \rangle \geq 0, \quad \text{Re} \langle \delta e - v, v - \gamma e \rangle \geq 0 \quad (3.181)$$

or equivalently,

$$\left\| u - \frac{\alpha + \beta}{2} e \right\| \leq \frac{1}{2} |\beta - \alpha|, \quad \left\| v - \frac{\gamma + \delta}{2} e \right\| \leq \frac{1}{2} |\delta - \gamma|, \quad (3.182)$$

then

$$|\langle u, v \rangle - \langle u, e \rangle \langle e, v \rangle| \leq \frac{1}{4} |\beta - \alpha| |\delta - \gamma|. \quad (3.183)$$

Applying (3.183) for $u = Ax$, $v = B^* x$ and $e = x$ we deduce

$$|\langle BAx, x \rangle - \langle Ax, x \rangle \langle Bx, x \rangle| \leq \frac{1}{4} |\beta - \alpha| |\delta - \gamma|, \quad (3.184)$$

for any $x \in H$, $\|x\| = 1$, which is an inequality of interest in itself.

Observing that

$$|\langle BAx, x \rangle| - |\langle Ax, x \rangle \langle Bx, x \rangle| \leq |\langle BAx, x \rangle - \langle Ax, x \rangle \langle Bx, x \rangle|,$$

then by (3.183) we deduce the inequality

$$|\langle BAx, x \rangle| \leq |\langle Ax, x \rangle \langle Bx, x \rangle| + \frac{1}{4} |\beta - \alpha| |\delta - \gamma|, \quad (3.185)$$

for any $x \in H$, $\|x\| = 1$. On taking the supremum over $\|x\| = 1$ in (3.185) we deduce the desired result (3.180). ∎

The following particular case provides a upper bound for the nonnegative quantity $\|A\|^2 - w(A)^2$ when some information about the operator A is available:

Corollary 174. *Let $A \in B(H)$ and $\alpha, \beta \in \mathbb{K}$ be such that the transform $C_{\alpha, \beta}(A)$ is accretive, then*

$$(0 \leq) \|A\|^2 - w^2(A) \leq \frac{1}{4} |\beta - \alpha|^2. \quad (3.186)$$

Proof. Follows on applying Theorem 173 above for the choice $B = A^*$, taking into account that $C_{\alpha, \beta}(A)$ is accretive implies that $C_{\bar{\alpha}, \bar{\beta}}(A^*)$ is the same and $w(A^* A) = \|A\|^2$. ∎

Remark 175. Let $A \in B(H)$ and $M > m > 0$ are such that the transform $C_{m,M}(A) = (A^* - mI)(MI - A)$ is accretive. Then

$$(0 \le) \|A\|^2 - w^2(A) \le \frac{1}{4}(M - m)^2. \tag{3.187}$$

A sufficient simple condition for $C_{m,M}(A)$ to be accretive is that A is a self-adjoint operator on H and such that $MI \ge A \ge mI$ in the partial operator order of $B(H)$.

The following result may be stated as well:

Theorem 176 (Dragomir [14], 2008). *Let $A, B \in B(H)$ and $\alpha, \beta, \gamma, \delta \in \mathbb{K}$ be such that* $\text{Re}(\beta\bar{\alpha}) > 0, \text{Re}(\delta\bar{\gamma}) > 0$ *and the transforms $C_{\alpha,\beta}(A), C_{\gamma,\delta}(B)$ are accretive, then*

$$\frac{w(BA)}{w(A)w(B)} \le 1 + \frac{1}{4} \cdot \frac{|\beta - \alpha||\delta - \gamma|}{[\text{Re}(\beta\bar{\alpha})\text{Re}(\delta\bar{\gamma})]^{\frac{1}{2}}} \tag{3.188}$$

and

$$w(BA) \le w(A)w(B) + \left[\left(|\alpha + \beta| - 2[\text{Re}(\beta\bar{\alpha})]^{\frac{1}{2}}\right)\right.$$
$$\left. \times \left(|\delta + \gamma| - 2[\text{Re}(\delta\bar{\gamma})]^{\frac{1}{2}}\right)\right]^{\frac{1}{2}} \times [w(A)w(B)]^{\frac{1}{2}}, \tag{3.189}$$

respectively.

Proof. With the assumptions (3.181) (or, equivalently, (3.182)) in the proof of Theorem 173) and if $\text{Re}(\beta\bar{\alpha}) > 0, \text{Re}(\delta\bar{\gamma}) > 0$, then

$$|\langle u, v \rangle - \langle u, e \rangle \langle e, v \rangle|$$
$$\le \begin{cases} \frac{1}{4}\frac{|\beta - \alpha||\delta - \gamma|}{[\text{Re}(\beta\bar{\alpha})\text{Re}(\delta\bar{\gamma})]^{\frac{1}{2}}}|\langle u, e \rangle \langle e, v \rangle|, \\ \left[\left(|\alpha + \beta| - 2[\text{Re}(\beta\bar{\alpha})]^{\frac{1}{2}}\right)\left(|\delta + \gamma| - 2[\text{Re}(\delta\bar{\gamma})]^{\frac{1}{2}}\right)\right]^{\frac{1}{2}} \\ \qquad\qquad\qquad\qquad \times [|\langle u, e \rangle \langle e, v \rangle|]^{\frac{1}{2}}. \end{cases} \tag{3.190}$$

The first inequality has been established in [4] (see [6, p. 62]) while the second one can be obtained in a canonical manner from the reverse of the Schwarz inequality given in [10]. The details are omitted.

Applying (3.183) for $u = Ax, v = B^*x$ and $e = x$ we deduce

$$|\langle BAx, x \rangle - \langle Ax, x \rangle \langle Bx, x \rangle|$$
$$\le \begin{cases} \frac{1}{4}\frac{|\beta - \alpha||\delta - \gamma|}{[\text{Re}(\beta\bar{\alpha})\text{Re}(\delta\bar{\gamma})]^{\frac{1}{2}}}|\langle A, x \rangle \langle Bx, x \rangle|, \\ \left[\left(|\alpha + \beta| - 2[\text{Re}(\beta\bar{\alpha})]^{\frac{1}{2}}\right)\left(|\delta + \gamma| - 2[\text{Re}(\delta\bar{\gamma})]^{\frac{1}{2}}\right)\right]^{\frac{1}{2}} \\ \qquad\qquad\qquad\qquad \times [|\langle A, x \rangle \langle Bx, x \rangle|]^{\frac{1}{2}} \end{cases} \tag{3.191}$$

for any $x \in H, \|x\| = 1$, which are of interest in themselves.

A similar argument to that in the proof of Theorem 173 yields the desired inequalities (3.188) and (3.189). The details are omitted.   ∎

Corollary 177. *Let $A \in B(H)$ and $\alpha, \beta \in \mathbb{K}$ be such that $\mathrm{Re}(\beta\bar{\alpha}) > 0$ and the transform $C_{\alpha,\beta}(A)$ is accretive, then*

$$(1 \leq) \frac{\|A\|^2}{w^2(A)} \leq 1 + \frac{1}{4} \cdot \frac{|\beta - \alpha|^2}{\mathrm{Re}(\beta\bar{\alpha})} \tag{3.192}$$

and

$$(0 \leq) \|A\|^2 - w^2(A) \leq \left(|\alpha + \beta| - 2[\mathrm{Re}(\beta\bar{\alpha})]^{\frac{1}{2}}\right) w(A) \tag{3.193}$$

respectively.

The proof is obvious from Theorem 176 on choosing $B = A^*$ and the details are omitted.

Remark 178. Let $A \in B(H)$ and $M > m > 0$ are such that the transform $C_{m,M}(A) = (A^* - mI)(MI - A)$ is accretive. Then, on making use of Corollary 177, we may state the following simpler results

$$(1 \leq) \frac{\|A\|}{w(A)} \leq \frac{1}{2} \cdot \frac{M + m}{\sqrt{Mm}} \tag{3.194}$$

and

$$(0 \leq) \|A\|^2 - w^2(A) \leq \left(\sqrt{M} - \sqrt{m}\right)^2 w(A) \tag{3.195}$$

respectively. These two inequalities were obtained earlier by the author using a different approach, see [13].

3.5.3 Some Particular Cases of Interest

The following result is well known in the literature (see for instance [25]):

$$w(A^n) \leq w^n(A),$$

for each positive integer n and any operator $A \in B(H)$.

The following reverse inequalities for $n = 2$, can be stated:

Proposition 179. *Let $A \in B(H)$ and $\alpha, \beta \in \mathbb{K}$ be such that the transform $C_{\alpha,\beta}(A)$ is accretive, then*

$$(0 \le) w^2(A) - w(A^2) \le \frac{1}{4} |\beta - \alpha|^2 . \tag{3.196}$$

Proof. On applying the inequality (3.184) from Theorem 173 for the choice $B = A$, we get the following inequality of interest in itself:

$$\left| \langle Ax, x \rangle^2 - \langle A^2 x, x \rangle \right| \le \frac{1}{4} |\beta - \alpha|^2 , \tag{3.197}$$

for any $x \in H, \|x\| = 1$. Since obviously,

$$|\langle Ax, x \rangle|^2 - \left| \langle A^2 x, x \rangle \right| \le \left| \langle Ax, x \rangle^2 - \langle A^2 x, x \rangle \right| ,$$

then by (3.197) we get

$$|\langle Ax, x \rangle|^2 \le \left| \langle A^2 x, x \rangle \right| + \frac{1}{4} |\beta - \alpha|^2 , \tag{3.198}$$

for any $x \in H, \|x\| = 1$. Taking the supremum over $\|x\| = 1$ in (3.198) we deduce the desired result (3.196). ∎

Remark 180. Let $A \in B(H)$ and $M > m > 0$ are such that the transform $C_{m,M}(A) = (A^* - mI)(MI - A)$ is accretive. Then

$$(0 \le) w^2(A) - w(A^2) \le \frac{1}{4}(M - m)^2 . \tag{3.199}$$

If $MI \ge A \ge mI$ in the partial operator order of $B(H)$, then (3.199) is valid.

Finally, we also have

Proposition 181. *Let $A \in B(H)$ and $\alpha, \beta \in \mathbb{K}$ be such that $\mathrm{Re}(\beta\overline{\alpha}) > 0$ and the transform $C_{\alpha,\beta}(A)$ is accretive, then*

$$(1 \le) \frac{w^2(A)}{w(A^2)} \le 1 + \frac{1}{4} \cdot \frac{|\beta - \alpha|^2}{\mathrm{Re}(\beta\overline{\alpha})} \tag{3.200}$$

and

$$(0 \le) w^2(A) - w(A^2) \le \left(|\alpha + \beta| - 2 \left[\mathrm{Re}(\beta\overline{\alpha}) \right]^{\frac{1}{2}} \right) w(A) \tag{3.201}$$

respectively.

Proof. On applying the inequality (3.191) from Theorem 176 for the choice $B = A$, we get the following inequality of interest in itself:

$$\left|\langle Ax, x\rangle^2 - \langle A^2 x, x\rangle\right| \leq \begin{cases} \frac{1}{4} \cdot \frac{|\beta - \alpha|^2}{\mathrm{Re}(\beta\overline{\alpha})} \, |\langle A, x\rangle|^2 , \\[2mm] \left(|\alpha + \beta| - 2 \left[\mathrm{Re}\,(\beta\overline{\alpha})\right]^{\frac{1}{2}}\right) |\langle A, x\rangle| . \end{cases} \qquad (3.202)$$

for any $x \in H, \|x\| = 1$.

Now, on making use of a similar argument to the one in the proof of Proposition 179, we deduce the desired results (3.200) and (3.201). The details are omitted. ∎

Remark 182. Let $A \in B(H)$ and $M > m > 0$ are such that the transform $C_{m,M}(A) = (A^* - mI)(MI - A)$ is accretive. Then, on making use of Proposition 181 we may state the following simpler results

$$(1 \leq)\, \frac{w^2(A)}{w(A^2)} \leq \frac{1}{4} \cdot \frac{(M+m)^2}{Mm} \qquad (3.203)$$

and

$$(0 \leq)\, w^2(A) - w(A^2) \leq \left(\sqrt{M} - \sqrt{m}\right)^2 w(A) \qquad (3.204)$$

respectively.

3.6   Some Inequalities for the Euclidean Operator Radius

3.6.1   Preliminary Facts

Let $B(H)$ denote the C^*−algebra of all bounded linear operators on a complex Hilbert space H with inner product $\langle \cdot, \cdot \rangle$. For $A \in B(H)$, let $w(A)$ and $\|A\|$ denote the numerical radius and the usual operator norm of A, respectively. It is well known that $w(\cdot)$ defines a norm on $B(H)$, and for every $A \in B(H)$,

$$\frac{1}{2}\|A\| \leq w(A) \leq \|A\| . \qquad (3.205)$$

For other results concerning the numerical range and radius of bounded linear operators on a Hilbert space, see [20] and [21].

In [24], F. Kittaneh has improved (3.205) in the following manner:

$$\frac{1}{4}\|A^*A + AA^*\| \leq w^2(A) \leq \frac{1}{2}\|A^*A + AA^*\| , \qquad (3.206)$$

with the constants $\frac{1}{4}$ and $\frac{1}{2}$ as best possible.

Following Popescu's work [26], we consider the *Euclidean operator radius* of a pair (C, D) of bounded linear operators defined on a Hilbert space $(H; \langle \cdot, \cdot \rangle)$. Note that in [26], the author has introduced the concept for an $n-$tuple of operators and pointed out its main properties.

Let (C, D) be a pair of bounded linear operators on H. The *Euclidean operator radius* is defined by:

$$w_e\,(C, D) := \sup_{\|x\|=1} \left(|\langle Cx, x\rangle|^2 + |\langle Dx, x\rangle|^2 \right)^{\frac{1}{2}}. \qquad (3.207)$$

As pointed out in [26], $w_e : B^2\,(H) \to [0, \infty)$ is a norm and the following inequality holds:

$$\frac{\sqrt{2}}{4} \|C^*C + D^*D\|^{\frac{1}{2}} \le w_e\,(C, D) \le \|C^*C + D^*D\|^{\frac{1}{2}}, \qquad (3.208)$$

where the constants $\frac{\sqrt{2}}{4}$ and 1 are best possible in (3.208).

We observe that if C and D are self-adjoint operators, then (3.208) becomes

$$\frac{\sqrt{2}}{4} \|C^2 + D^2\|^{\frac{1}{2}} \le w_e\,(C, D) \le \|C^2 + D^2\|^{\frac{1}{2}}. \qquad (3.209)$$

We observe also that if $A \in B\,(H)$ and $A = B + iC$ is the Cartesian decomposition of A, then

$$w_e^2\,(B, C) = \sup_{\|x\|=1} \left[|\langle Bx, x\rangle|^2 + |\langle Cx, x\rangle|^2 \right]$$

$$= \sup_{\|x\|=1} |\langle Ax, x\rangle|^2 = w^2\,(A).$$

By the inequality (3.209) and since (see [24])

$$A^*A + AA^* = 2\left(B^2 + C^2\right), \qquad (3.210)$$

then we have

$$\frac{1}{16} \|A^*A + AA^*\| \le w^2\,(A) \le \frac{1}{2} \|A^*A + AA^*\|. \qquad (3.211)$$

We remark that the lower bound for $w^2\,(A)$ in (3.211) provided by Popescu's inequality (3.208) is not as good as the first inequality of Kittaneh from (3.206). However, the upper bounds for $w^2\,(A)$ are the same and have been proved using different arguments.

The main aim of this section is to extend Kittaneh's result to the Euclidean radius of two operators and investigate other particular instances of interest. Related results connecting the Euclidean operator radius, the usual numerical radius of a composite operator and the operator norm are also provided.

3.6.2 Some Inequalities for the Euclidean Operator Radius

The following result concerning a sharp lower bound for the Euclidean operator radius may be stated:

Theorem 183 (Dragomir [8], 2006). *Let* $B, C : H \to H$ *be two bounded linear operators on the Hilbert space* $(H; \langle \cdot, \cdot \rangle)$ *. Then*

$$\frac{\sqrt{2}}{2} \left[w \left(B^2 + C^2 \right) \right]^{\frac{1}{2}} \leq w_e (B, C) \left(\leq \| B^* B + C^* C \|^{\frac{1}{2}} \right). \qquad (3.212)$$

The constant $\frac{\sqrt{2}}{2}$ *is best possible in the sense that it cannot be replaced by a larger constant.*

Proof. We follow a similar argument to the one from [24].
  For any $x \in H$, $\|x\| = 1$, we have

$$|\langle Bx, x \rangle|^2 + |\langle Cx, x \rangle|^2 \geq \frac{1}{2} \left(|\langle Bx, x \rangle| + |\langle Cx, x \rangle| \right)^2$$

$$\geq \frac{1}{2} |\langle (B \pm C) x, x \rangle|^2. \qquad (3.213)$$

Taking the supremum in (3.213), we deduce

$$w_e^2 (B, C) \geq \frac{1}{2} w^2 (B \pm C). \qquad (3.214)$$

Utilizing the inequality (3.214) and the properties of the numerical radius, we have successively:

$$2w_e^2 (B, C) \geq \frac{1}{2} \left[w^2 (B + C) + w^2 (B - C) \right]$$

$$\geq \frac{1}{2} \left\{ w \left[(B + C)^2 \right] + w \left[(B - C)^2 \right] \right\}$$

$$\geq \frac{1}{2} \left\{ w \left[(B + C)^2 + (B - C)^2 \right] \right\}$$

$$= w \left(B^2 + C^2 \right),$$

which gives the desired inequality (3.212).
  The sharpness of the constant will be shown in a particular case, later on. ∎

Corollary 184. *For any two self-adjoint bounded linear operators* B, C *on* H, *we have*

$$\frac{\sqrt{2}}{2} \left\| B^2 + C^2 \right\|^{\frac{1}{2}} \le w_e\left(B, C\right) \left(\le \left\| B^2 + C^2 \right\|^{\frac{1}{2}}\right). \tag{3.215}$$

The constant $\frac{\sqrt{2}}{2}$ is sharp in (3.215).

Remark 185. The inequality (3.215) is better than the first inequality in (3.209) which follows from Popescu's first inequality in (3.208). It also provides, for the case that B, C are the self-adjoint operators in the Cartesian decomposition of A, exactly the lower bound obtained by Kittaneh in (3.206) for the numerical radius $w\left(A\right)$. Moreover, since $\frac{1}{4}$ is a sharp constant in Kittaneh's inequality (3.206), it follows that $\frac{\sqrt{2}}{2}$ is also the best possible constant in (3.215) and (3.212), respectively.

The following particular case may be of interest:

Corollary 186. *For any bounded linear operator $A : H \rightarrow H$ and $\alpha, \beta \in \mathbb{C}$ we have*

$$\frac{1}{2} w\left[\alpha^2 A^2 + \beta^2 \left(A^*\right)^2\right] \le \left(|\alpha|^2 + |\beta|^2\right) w^2\left(A\right)$$

$$\left(\le \left\| |\alpha|^2 A^* A + |\beta|^2 A A^* \right\|\right). \tag{3.216}$$

Proof. If we choose in Theorem 183, $B = \alpha A$ and $C = \beta A^*$, we get

$$w_e^2\left(B, C\right) = \left(|\alpha|^2 + |\beta|^2\right) w^2\left(A\right)$$

and

$$w\left(B^2 + C^2\right) = w\left[\alpha^2 A^2 + \beta^2 \left(A^*\right)^2\right],$$

which by (3.212) implies the desired result (3.216).   ∎

Remark 187. If we choose in (3.216) $\alpha = \beta \ne 0$, then we get the inequality

$$\frac{1}{4} \left\| A^2 + \left(A^*\right)^2 \right\| \le w^2\left(A\right) \left(\le \frac{1}{2} \left\| A^* A + A A^* \right\|\right), \tag{3.217}$$

for any bounded linear operator $A \in B\left(H\right)$.

If we choose in (3.216), $\alpha = 1, \beta = i$, then we get

$$\frac{1}{4} w\left[A^2 - \left(A^*\right)^2\right] \le w^2\left(A\right), \tag{3.218}$$

for every bounded linear operator $A : H \rightarrow H$.

The following result may be stated as well.

Theorem 188 (Dragomir [8], 2006). *For any two bounded linear operators B, C on H we have*

$$\frac{\sqrt{2}}{2} \max \{w(B+C), w(B-C)\} \leq w_e(B, C) \tag{3.219}$$

$$\leq \frac{\sqrt{2}}{2} \left[w^2(B+C) + w^2(B-C) \right]^{\frac{1}{2}}.$$

The constant $\frac{\sqrt{2}}{2}$ is sharp in both inequalities.

Proof. The first inequality follows from (3.214).

For the second inequality, we observe that

$$|\langle Cx, x \rangle \pm \langle Bx, x \rangle|^2 \leq w^2(C \pm B) \tag{3.220}$$

for any $x \in H$, $\|x\| = 1$.

The inequality (3.220) and the parallelogram identity for complex numbers give:

$$2 \left[|\langle Bx, x \rangle|^2 + |\langle Cx, x \rangle|^2 \right]$$

$$= |\langle Bx, x \rangle - \langle Cx, x \rangle|^2 + |\langle Bx, x \rangle + \langle Cx, x \rangle|^2$$

$$\leq w^2(B+C) + w^2(B-C), \tag{3.221}$$

for any $x \in H$, $\|x\| = 1$.

Taking the supremum in (3.220) we deduce the desired result (3.219).

The fact that $\frac{\sqrt{2}}{2}$ is the best possible constant follows from the fact that for $B = C \neq 0$ one would obtain the same quantity $\sqrt{2} w(B)$ in all terms of (3.219). ∎

Corollary 189. *For any two self-adjoint operators B, C on H we have*

$$\frac{\sqrt{2}}{2} \max \{ \|B+C\|, \|B-C\| \}$$

$$\leq w_e(B, C) \leq \frac{\sqrt{2}}{2} \left[\|B+C\|^2 + \|B-C\|^2 \right]^{\frac{1}{2}}. \tag{3.222}$$

The constant $\frac{\sqrt{2}}{2}$ is best possible in both inequalities.

Corollary 190. *Let A be a bounded linear operator on H. Then*

$$\frac{\sqrt{2}}{2} \max \left\{ \left\| \frac{(1-i)A + (1+i)A^*}{2} \right\|, \left\| \frac{(1+i)A + (1-i)A^*}{2} \right\| \right\}$$

$$\leq w(A)$$

$$\leq \frac{\sqrt{2}}{2} \left[\left\| \frac{(1-i)\,A + (1+i)\,A^*}{2} \right\|^2 + \left\| \frac{(1+i)\,A + (1-i)\,A^*}{2} \right\|^2 \right]^{\frac{1}{2}}. \quad (3.223)$$

Proof. Follows from (3.222) applied for the Cartesian decomposition of A.   ∎

The following result may be stated as well:

Corollary 191. *For any A a bounded linear operator on H and $\alpha, \beta \in \mathbb{C}$, we have*

$$\frac{\sqrt{2}}{2} \max \left\{ w\left(\alpha A + \beta A^*\right), w\left(\alpha A - \beta A^*\right) \right\}$$

$$\leq \left(|\alpha|^2 + |\beta|^2 \right)^{\frac{1}{2}} w\left(A\right)$$

$$\leq \frac{\sqrt{2}}{2} \left[w^2 \left(\alpha A + \beta A^*\right) + w^2 \left(\alpha A - \beta A^*\right) \right]^{\frac{1}{2}}. \quad (3.224)$$

Remark 192. The above inequality (3.224) contains some particular cases of interest. For instance, if $\alpha = \beta \neq 0$, then by (3.224) we get

$$\frac{1}{2} \max \left\{ \|A + A^*\|, \|A - A^*\| \right\}$$

$$\leq w\left(A\right) \leq \frac{1}{2} \left[\|A + A^*\|^2 + \|A - A^*\|^2 \right]^{\frac{1}{2}}, \quad (3.225)$$

since, obviously $w\left(A + A^*\right) = \|A + A^*\|$ and $w\left(A - A^*\right) = \|A - A^*\|$, $A - A^*$ being a normal operator.

Now, if we choose in (3.224), $\alpha = 1$ and $\beta = i$, and taking into account that $A + iA^*$ and $A - iA^*$ are normal operators, then we get

$$\frac{1}{2} \max \left\{ \|A + iA^*\|, \|A - iA^*\| \right\}$$

$$\leq w\left(A\right) \leq \frac{1}{2} \left[\|A + iA^*\|^2 + \|A - iA^*\|^2 \right]^{\frac{1}{2}}. \quad (3.226)$$

The constant $\frac{1}{2}$ is best possible in both inequalities (3.225) and (3.226).

The following simple result may be stated as well.

Proposition 193 (Dragomir [8], 2006). *For any two bounded linear operators B and C on H, we have the inequality:*

$$w_e\left(B, C\right) \leq \left[w^2 \left(C - B\right) + 2w\left(B\right) w\left(C\right) \right]^{\frac{1}{2}}. \quad (3.227)$$

Proof. For any $x \in H$, $\|x\| = 1$, we have

$$|\langle Cx, x \rangle|^2 - 2\,\mathrm{Re}\left[\langle Cx, x \rangle \,\overline{\langle Bx, x \rangle}\right] + |\langle Bx, x \rangle|^2$$

$$= |\langle Cx, x \rangle - \langle Bx, x \rangle|^2 \leq w^2 (C - B), \qquad (3.228)$$

giving

$$|\langle Cx, x \rangle|^2 + |\langle Bx, x \rangle|^2 \leq w^2 (C - B) + 2\,\mathrm{Re}\left[\langle Cx, x \rangle \,\overline{\langle Bx, x \rangle}\right]$$

$$\leq w^2 (C - B) + 2\,|\langle Cx, x \rangle|\,|\langle Bx, x \rangle| \qquad (3.229)$$

for any $x \in H$, $\|x\| = 1$.

Taking the supremum in (3.229) over $\|x\| = 1$, we deduce the desired inequality (3.227). ∎

In particular, if B and C are self-adjoint operators, then

$$w_e (B, C) \leq \left(\|B - C\|^2 + 2\,\|B\|\,\|C\|\right)^{\frac{1}{2}}. \qquad (3.230)$$

Now, if we apply the inequality (3.230) for $B = \frac{A + A^*}{2}$ and $C = \frac{A - A^*}{2i}$, where $A \in B(H)$, then we deduce

$$w(A) \leq \left[\left\|\frac{(1+i) A + (1-i) A^*}{2}\right\|^2 + 2 \cdot \left\|\frac{A + A^*}{2}\right\|\left\|\frac{A - A^*}{2}\right\|\right]^{\frac{1}{2}}.$$

The following result provides a different upper bound for the Euclidean operator radius than (3.227).

Proposition 194 (Dragomir [8], 2006). *For any two bounded linear operators B and C on H, we have*

$$w_e (B, C) \leq \left[2 \min \left\{w^2 (B), w^2 (C)\right\} + w (B - C) w (B + C)\right]^{\frac{1}{2}}. \qquad (3.231)$$

Proof. Utilizing the parallelogram identity (3.221), we have, by taking the supremum over $x \in H, \|x\| = 1$, that

$$2w_e^2 (B, C) = w_e^2 (B - C, B + C). \qquad (3.232)$$

Now, if we apply Proposition 193 for $B - C, B + C$ instead of B and C, then we can state

$$w_e^2 (B - C, B + C) \leq 4w^2 (C) + 2w (B - C) w (B + C)$$

giving

$$w_e^2(B,C) \leq 2w^2(C) + w(B-C)\,w(B+C). \tag{3.233}$$

Now, if in (3.233) we swap the C with B, then we also have

$$w_e^2(B,C) \leq 2w^2(B) + w(B-C)\,w(B+C). \tag{3.234}$$

The conclusion follows now by (3.233) and (3.234).   ∎

3.6.3   Other Results

A different upper bound for the Euclidean operator radius is incorporated in the following

Theorem 195 (Dragomir [8], 2006). *Let* $(H; \langle \cdot, \cdot \rangle)$ *be a Hilbert space and* B, C *two bounded linear operators on* H. *Then*

$$w_e^2(B,C) \leq \max\left\{\|B\|^2, \|C\|^2\right\} + w\left(C^*B\right). \tag{3.235}$$

The inequality (3.235) is sharp.

Proof. Firstly, let us observe that for any $y, u, v \in H$, we have successively

$$\|\langle y, u\rangle u + \langle y, v\rangle v\|^2$$

$$= |\langle y, u\rangle|^2 \|u\|^2 + |\langle y, v\rangle|^2 \|v\|^2 + 2\operatorname{Re}\left[\langle y, u\rangle \overline{\langle y, v\rangle} \langle u, v\rangle\right]$$

$$\leq |\langle y, u\rangle|^2 \|u\|^2 + |\langle y, v\rangle|^2 \|v\|^2 + 2|\langle y, u\rangle| |\langle y, v\rangle| |\langle u, v\rangle|$$

$$\leq |\langle y, u\rangle|^2 \|u\|^2 + |\langle y, v\rangle|^2 \|v\|^2 + \left(|\langle y, u\rangle|^2 + |\langle y, v\rangle|^2\right) |\langle u, v\rangle|$$

$$\leq \left(|\langle y, u\rangle|^2 + |\langle y, v\rangle|^2\right) \left(\max\left\{\|u\|^2, \|v\|^2\right\} + |\langle u, v\rangle|\right). \tag{3.236}$$

On the other hand,

$$\left(|\langle y, u\rangle|^2 + |\langle y, v\rangle|^2\right)^2 = [\langle y, u\rangle \langle u, y\rangle + \langle y, v\rangle \langle v, y\rangle]^2$$

$$= [\langle y, \langle y, u\rangle u + \langle y, v\rangle v\rangle]^2$$

$$\leq \|y\|^2 \|\langle y, u\rangle u + \langle y, v\rangle v\|^2 \tag{3.237}$$

for any $y, u, v \in H$.

Making use of (3.236) and (3.237) we deduce that

$$|\langle y, u\rangle|^2 + |\langle y, v\rangle|^2 \leq \|y\|^2 \left[\max\left\{\|u\|^2, \|v\|^2\right\} + |\langle u, v\rangle|\right] \tag{3.238}$$

for any $y, u, v \in H$, which is a vector inequality of interest in itself.

Now, if we apply the inequality (3.238) for $y = x, u = Bx, v = Cx, x \in H$, $\|x\| = 1$, then we can state that

$$|\langle Bx, x \rangle|^2 + |\langle Cx, x \rangle|^2 \leq \max \left\{ \|Bx\|^2, \|Cx\|^2 \right\} + |\langle Bx, Cx \rangle| \qquad (3.239)$$

for any $x \in H$, $\|x\| = 1$, which is of interest in itself.

Taking the supremum over $x \in H$, $\|x\| = 1$, we deduce the desired result (3.235).

To prove the sharpness of the inequality (3.235) we choose $C = B$, B a self-adjoint operator on H. In this case, both sides of (3.235) become $2\|B\|^2$. ∎

If information about the sum and the difference of the operators B and C is available, then one may use the following result:

Corollary 196. *For any two operators $B, C \in B(H)$ we have*

$$w_e^2 (B, C)$$
$$\leq \frac{1}{2} \left\{ \max \left\{ \|B - C\|^2, \|B + C\|^2 \right\} + w \left[(B^* - C^*)(B + C) \right] \right\}. \qquad (3.240)$$

The constant $\frac{1}{2}$ is best possible in (3.240).

Proof. Follows by the inequality (3.235) written for $B + C$ and $B - C$ instead of B and C and by utilizing the identity (3.232).

The fact that $\frac{1}{2}$ is best possible in (3.240) follows by the fact that for $C = B$, B a self-adjoint operator, we get in both sides of the inequality (3.240) the quantity $2\|B\|^2$. ∎

Corollary 197. *Let $A : H \to H$ be a bounded linear operator on the Hilbert space H. Then*

$$w^2 (A)$$
$$\leq \frac{1}{4} \left[\max \left\{ \|A + A^*\|^2, \|A - A^*\|^2 \right\} + w \left[(A^* - A)(A + A^*) \right] \right]. \qquad (3.241)$$

The constant $\frac{1}{4}$ is best possible.

Proof. If $B = \dfrac{A + A^*}{2}, C = \dfrac{A - A^*}{2i}$ is the Cartesian decomposition of A, then

$$w_e^2 (B, C) = w^2 (A)$$

and

$$w (C^* B) = \frac{1}{4} w \left[(A^* - A)(A + A^*) \right].$$

Utilizing (3.235) we deduce (3.241).   ∎

Remark 198. If we choose in (3.235), $B = A$ and $C = A^*$, $A \in B(H)$, then we can state that

$$w^2(A) \le \frac{1}{2}\left[\|A\|^2 + w\left(A^2\right)\right]. \tag{3.242}$$

The constant $\frac{1}{2}$ is best possible in (3.242).

Note that this inequality has been obtained in [11] by the use of a different argument based on the Buzano's inequality.

Finally, the following upper bound for the Euclidean radius involving different composite operators also holds:

Theorem 199 (Dragomir [8], 2006). *With the assumptions of Theorem 195, we have*

$$w_e^2(B,C) \le \frac{1}{2}\left[\|B^*B + C^*C\| + \|B^*B - C^*C\|\right] + w\left(C^*B\right). \tag{3.243}$$

The inequality (3.243) is sharp.

Proof. We use (3.239) to write that

$$|\langle Bx, x\rangle|^2 + |\langle Cx, x\rangle|^2$$
$$\le \frac{1}{2}\left[\|Bx\|^2 + \|Cx\|^2 + \left|\|Bx\|^2 - \|Cx\|^2\right|\right] + |\langle Bx, Cx\rangle| \tag{3.244}$$

for any $x \in H, \|x\| = 1$.

Since $\|Bx\|^2 = \langle B^*Bx, x\rangle$, $\|Cx\|^2 = \langle C^*Cx, x\rangle$, then (3.244) can be written as

$$|\langle Bx, x\rangle|^2 + |\langle Cx, x\rangle|^2$$
$$\le \frac{1}{2}\left[\langle(B^*B + C^*C)x, x\rangle + |\langle(B^*B - C^*C)x, x\rangle|\right] + |\langle Bx, Cx\rangle| \tag{3.245}$$

$x \in H, \|x\| = 1$.

Taking the supremum in (3.245) over $x \in H, \|x\| = 1$ and noting that the operators $B^*B \pm C^*C$ are self-adjoint, we deduce the desired result (3.243).

The sharpness of the constant will follow from that of (3.248) pointed out below.   ∎

Corollary 200. *For any two operators $B, C \in B(H)$, we have*

$$w_e^2(B,C)$$
$$\le \frac{1}{2}\left\{\|B^*B + C^*C\| + \|B^*C + C^*B\| + w\left[(B^* - C^*)(B + C)\right]\right\}. \tag{3.246}$$

The constant $\frac{1}{2}$ is best possible.

Proof. If we write (3.243) for $B + C, B - C$ instead of B, C and perform the required calculations, then we get

$$w_e^2 (B + C, B - C)$$
$$\leq \|B^* B + C^* C\| + \|B^* C + C^* B\| + w\left[(B^* - C^*)(B + C)\right],$$

which, by the identity (3.232) is clearly equivalent with (3.246).

Now, if we choose in (3.246) $B = C$, then we get the inequality $w(B) \leq \|B\|$, which is a sharp inequality. ∎

Corollary 201. *If B, C are self-adjoint operators on H, then*

$$w_e^2 (B, C) \leq \frac{1}{2}\left[\|B^2 + C^2\| + \|B^2 - C^2\|\right] + w(CB). \tag{3.247}$$

We observe that, if B and C are chosen to be the Cartesian decomposition for the bounded linear operator A, then we can get from (3.247) that

$$w^2 (A)$$
$$\leq \frac{1}{4}\left\{\|A^* A + AA^*\| + \|A^2 + (A^*)^2\| + w\left[(A^* - A)(A + A^*)\right]\right\}. \tag{3.248}$$

The constant $\frac{1}{4}$ is best possible. This follows by the fact that for A a self-adjoint operator, we obtain on both sides of (3.248) the same quantity $\|A\|^2$.

Now, if we choose in (3.243) $B = A$ and $C = A^*$, $A \in B(H)$, then we get

$$w^2 (A) \leq \frac{1}{4}\left\{\|A^* A + AA^*\| + \|A^* A - AA^*\|\right\} + \frac{1}{2}w(A^2). \tag{3.249}$$

This inequality is sharp. The equality holds if, for instance, we assume that A is normal, i.e., $A^* A = AA^*$. In this case we get on both sides of (3.249) the quantity $\|A\|^2$, since for normal operators, $w(A^2) = w^2(A) = \|A\|^2$.

References

1. Dragomir, S.S.: A generalisation of Grüss' inequality in inner product spaces and applications. J. Math. Anal. Applic. **237**, 74–82 (1999)
2. Dragomir, S.S.: Some Grüss type inequalities in inner product spaces. J. Inequal. Pure & Appl. Math. **4**(2), Article 42 (2003)
3. Dragomir, S.S.: New reverses of Schwarz, triangle and Bessel inequalities in inner product spaces. Australian J. Math. Anal. & Appl. **1**(1), Article 1, 1–18 (2004)

4. Dragomir, S.S.: Reverses of Schwarz, triangle and Bessel inequalities in inner product spaces. J. Inequal. Pure & Appl. Math., **5**(3), Article 76 (2004)
5. Dragomir, S.S.: A counterpart of Schwarz's inequality in inner product spaces. East Asian Math. J. **20**(1), 1–10 (2004)
6. Dragomir, S.S.: Advances in Inequalities of the Schwarz, Gruss and Bessel Type in Inner Product Spaces, Nova Science Publishers, Inc., New York (2005)
7. Dragomir, S.S.: Inequalities for the norm and the numerical radius of composite operators in Hilbert spaces. Inequalities and applications. Internat. Ser. Numer. Math. 157, 135–146 Birkhäuser, Basel, 2009. Preprint available in RGMIA Res. Rep. Coll. **8** (2005), Supplement, Article 11
8. Dragomir, S.S.: Some inequalities for the Euclidean operator radius of two operators in Hilbert spaces. Linear Algebra Appl. **419**(1), 256–264 (2006)
9. Dragomir, S.S.: Reverse inequalities for the numerical radius of linear operators in Hilbert spaces. Bull. Austral. Math. Soc. **73**(2), 255–262 (2006)
10. Dragomir, S.S.: Reverses of the Schwarz inequality generalising a Klamkin-McLenaghan result. Bull. Austral. Math. Soc.**73**(1), 69–78 (2006)
11. Dragomir, S.S.: Inequalities for the norm and the numerical radius of linear operators in Hilbert spaces. Demonstratio Math. **40**(2), 411–417 (2007)
12. Dragomir, S.S.: Norm and numerical radius inequalities for a product of two linear operators in Hilbert spaces. J. Math. Inequal. **2**(4), 499–510 (2008)
13. Dragomir, S.S.: New inequalities of the Kantorovich type for bounded linear operators in Hilbert spaces. Linear Algebra Appl. **428**(11–12), 2750–2760 (2008)
14. Dragomir, S.S.: Some inequalities of the Grüss type for the numerical radius of bounded linear operators in Hilbert spaces. J. Inequal. Appl. **2008**, Art. ID 763102, 9 pp. Preprint, RGMIA Res. Rep. Coll. **11**(1) (2008)
15. Dragomir, S.S.: Some inequalities for commutators of bounded linear operators in Hilbert spaces, Preprint. RGMIA Res. Rep. Coll.**11**(1), Article 7 (2008)
16. Dragomir, S.S.: Power inequalities for the numerical radius of a product of two operators in Hilbert spaces.Sarajevo J. Math. **5**(18)(2), 269–278 (2009)
17. Dragomir, S.S.: A functional associated with two bounded linear operators in Hilbert spaces and related inequalities. Ital. J. Pure Appl. Math. No. **27**, 225–240 (2010)
18. Dragomir, S.S., SÁNDOR, J.: Some inequalities in prehilbertian spaces. Studia Univ. "Babeş-Bolyai" - Mathematica **32**(1), 71–78 (1987)
19. Goldstein, A., Ryff, J.V., Clarke, L.E.: Problem 5473. Amer. Math. Monthly **75**(3), 309 (1968)
20. Gustafson, K.E., Rao, D.K.M. Numerical Range Springer, New York, Inc. (1997)
21. Halmos, P.R.: A Hilbert Space Problem Book, 2nd edn. Springer-Verlag, New York, Heidelberg, Berlin (1982)
22. Kittaneh, F.: Notes on some inequalities for Hilbert space operators. Publ. Res. Inst. Math. Sci. **24**, 283–293 (1988)
23. Kittaneh, F.: A numerical radius inequality and an estimate for the numerical radius of the Frobenius companion matrix. Studia Math. **158**(1), 11–17 (2003)
24. Kittaneh, F.: Numerical radius inequalities for Hilbert space operators. Studia Math. **168**(1), 73–80 (2005)
25. Pearcy, C.: An elementary proof of the power inequality for the numerical radius. Michigan Math. J. **13**, 289–291 (1966)
26. Popescu, G.: Unitary invariants in multivariable operator theory. Mem. Amer. Math. Soc. **200**(941), vi+91 (2009) ISBN: 978-0-8218-4396-3. Preprint, Arχiv.math.OA/0410492